John Herbert A. Bone

Petroleum and Petroleum Wells

With a Complete Guide Book and Description of the Oil Regions of Pennsylvania...

Second Edition

John Herbert A. Bone

Petroleum and Petroleum Wells
With a Complete Guide Book and Description of the Oil Regions of Pennsylvania... Second Edition

ISBN/EAN: 9783744758871

Printed in Europe, USA, Canada, Australia, Japan

Cover: Foto ©berggeist007 / pixelio.de

More available books at **www.hansebooks.com**

PETROLEUM

AND

PETROLEUM WELLS.

WHAT PETROLEUM IS, WHERE IT IS FOUND, AND WHAT IT
IS USED FOR; WHERE TO SINK PETROLEUM
WELLS, AND HOW TO SINK THEM.

WITH

A COMPLETE GUIDE BOOK

AND

DESCRIPTION OF THE OIL REGIONS

OF

PENNSYLVANIA, WEST VIRGINIA, KENTUCKY AND OHIO.

By J. H. A. BONE.

SECOND EDITION, REVISED AND ENLARGED.

PHILADELPHIA
J. B. LIPPINCOTT & CO.
1865.

INTRODUCTION.

In the present volume the author has attempted to supply a want widely felt and frequently expressed—that of a popular handbook on the subject of .petroleum and petroleum enterprises. So far as he is aware, there is no work now before the public that aims to give, what is here attempted, a clear and systematic account of the origin, description, and history of petroleum; its distribution over the globe; the recent discovery of its existence in the United States, with a complete history of petroleum enterprise in this country from its inception; the method of working, and a minute and accurate account of the different regions of its production, forming at once a history, a practical treatise, and a guide-book, embodying everything important to know in reference to petroleum and the oil regions, not only of Pennsylvania, but of the other oil producing States.

The facts contained in this volume have been obtained by thorough and careful personal observation, and as the author has no " axe to grind," every effort has been made to produce a reliable work. It may be proper to say, that some portions of the book have appeared in substance in journals with which the author is connected.

May, 1865.

CONTENTS.

(v)

vi CONTENTS.

PETROLEUM

AND

PETROLEUM WELLS.

PETROLEUM, ITS DESCRIPTION AND HISTORY.—HOW IT IS FORMED AND WHERE IT IS FOUND.

WHAT Petroleum is, where it is to be found, and what are the causes of its formation, are subjects now engaging the attention of the civilized world, and to neither of these questions have perfectly satisfactory answers yet been given. The name itself is from the Latin *petra*, a rock, and *oleum*, oil, being in fact "rock oil," deriving its name from being found in the rocks, or oozing from them. In its natural state its composition is very indefinite, consisting of various oily hydro-carbons, holding in solution paraffine and solid bitumen, or asphaltum. In some scientific works the fluid petroleum is described

(7)

under the name "naphtha oil," whilst that having a large proportion of asphaltum, is known as "bitumen." The latter is of comparatively little value, but the fluid petroleum, since the discovery of its manifold and important uses, has risen to be one of the most prominent staples. The lighter oil, cleansed and purified, has come into almost universal request as an illuminator, surpassing all others, except gas, in brilliancy, and also possessing the merit of cheapness. The secret of producing gas itself, equal in illuminating power to the best coal gas, produced with much greater ease and at less expense, has been discovered and put into practice; whilst, to show the capabilities of petroleum as an illuminator, the solid residuum of the refining process is made into paraffine candles. As a lubricator for wheels and machinery the heavier qualities of petroleum have come into general use. Paint oils and varnish are made from it, and the benzine is used as a substitute for turpentine. Petrolized soap is a favorite toilet article. The most beautiful and durable colors and shades now in wear are obtained from the waste petroleum after refining. It has been used with success as a substitute for fish oil in tanning. For generations it has proved a valuable medicine, applied both externally and internally. In fact, there seems to be no limit to its usefulness, for new applications of it are frequently discovered.

Petroleum, in one form or another, has been known in all ages, and in nearly all parts of the world, although many of its uses are the discoveries of the past few years. It is mentioned by the ancient Greeks and Romans, being known to the latter under the name of "bitumen." At Zante, one of the Ionian Islands, is an oil spring, still flowing, which was mentioned by Herodotus, more than two thousand years ago. In Sicily the ancient inhabitants burned petroleum in their lamps instead of fish oil. In the north of Italy it has for nearly two centuries furnished material for lighting the streets of Genoa and Parma. On the shores of the Caspian Sea, at Bakoo, are extraordinary manifestations of petroleum oil and gas. These extend over a tract of country about twenty-five miles in length, and about half a mile wide, in strata of a porous, argillaceous sandstone, belonging to the tertiary period. In the vicinity are hills of volcanic rocks, through which springs of the heavier sort of petroleum flow. Open wells, from sixteen to twenty feet deep, are dug, and in these the oil gathers as it oozes from the strata. A large amount is annually gathered and distributed over Persia, where it is exclusively used for illuminating purposes, and for the sacred fires. The Rangoon district, on the Irrawaddy, is also famous for its large product of rock oil, and for centuries the whole Burman empire has been supplied with oil from this source. The annual yield of petroleum from this district is said to be more than 400,000 hogsheads, or about two thirds of the export from New York for 1864. The number of wells is 520. The natives use the oil as a medicine, burn it in their lamps, and grease timber with it to prevent the destructive operations of insects. Some of the Burmese oil has been sent to England and used in the manufacture of paraffine candles. In consistency it resembles the heavy lubricating oils of Pennsylvania and Ohio, whilst its color, of a greenish brown, is more like that of the lighter Pennsylvania oil. Petroleum is frequently found in the neighborhood of volcanoes. Around the volcanic isles of Cape Verde it is seen floating on the water; and to the south of Vesuvius a spring of it rises through the sea.

But it is in America that the largest deposits of liquid petroleum are found. Besides the principal reservoirs in Northwestern Pennsylvania, there are other deposits, the full value of which have not yet been ascertained, in Southwestern Pennsylvania, Ohio, Western Virginia, Kentucky, New York, Canada, Kansas, and California. Indications of its existence have also been discovered in Michigan, Indiana, Illinois, and Iowa, and experiments are in progress to test the value and extent of the deposits.

The causes of formation of petroleum, and its location in the rocks, are questions that have as yet received no satisfactory solution. According to some geologists the oil originates in the coal beds, from which it is expelled by pressure, whilst others assert that the coal is formed from the oil, instead of the oil from the coal. In support of both of these theories the general resemblance of petroleum to the oil obtained from the distillation of coal is adduced, although there are some minor points of difference. But the existence of petroleum does not depend on the existence of coal in the same locality; on the contrary, the most productive oil districts are removed from the coal fields. In the Pennsylvania oil region the wells are entirely outside of the coal field, and so remote from it that there can scarcely be any connection between the oil and coal beds. The strata in which the oil is found dip south, and pass below the coal measures from five hundred to one thousand feet, the nearest coal bed to the more northern oil wells capping the highest hills about thirty miles distant.

Other geologists attribute the production of the oil to the slow distillation of animal or vegetable matter overwhelmed by ancient floods, and

imprisoned in the rocks formed from the sand or
mud in which the organic remains were buried.
This theory presupposes an immense deposit of
animal or vegetable matter, as the yield of oil
has already been very large, and but a small
portion of the deposit has been developed as yet.
Another theory accounts for its production by
volcanic agencies, but it is not by any means
confined to the volcanic rocks. Some are disposed to look on it as a formation of by-gone
ages, by processes long since terminated, whilst
others, with a belief in the doctrine that Nature
never stops in her work, assert that the process
of formation is still going on, and that the supply is inexhaustible. An apparent confirmation
of this opinion is found in the fact that the wells
of Bakoo and Rangoon are as productive now as
they were centuries ago. Single wells have dried
up, but new ones have been sunk, and the product of the district suffers no diminution. This
fact should allay the fears of those who are ap-
prehensive that the American oil regions will
soon be exhausted.

Petroleum is found in different parts of the
world in all the stratified rocks, and in the volcanic
and metamorphic formations. It is sometimes
traced to beds of lignite, and sometimes its source
cannot be discovered. In the United States and
Canada the sandstones are the most productive

of oil. In the Pennsylvania oil region the hills are capped with conglomerate, lying in geological succession next below the coal measures. Through this the well is bored, passing through alternating layers of shale and sandstone, and terminating in sandstone, where cavities exist, frequently filled with oil, gas, and salt water. The dip of the strata in Northwestern Pennsylvania is nearly south. In Ohio it is east of south. The most productive oil bearing sandstone crops out in Ohio a few miles west of Cleveland, and dips gently towards the Alleghany River, descending more rapidly as it gets farther south. In some parts of Oil Creek, and on the Alleghany, there are appearances of a slight upheaval, forming cracks and fissures in the rocks, and it is here that many borers look most hopefully for oil in large quantities. According to Prof. Evans, of Marietta, who has given the matter much study, the oil is contained in cavities or fissures in the rocks, in connection with both water and gas. These are arranged, of course, according to their weight, the water at the bottom, the oil floating thereon, and the gas (often strongly compressed) filling the upper part of the cavity. If such a cavity runs obliquely from above downward, a well, when bored, may strike either the water or the oil, or it may enter the gas chambers. In the first two cases, if the gas be com-

2

pressed, as it usually is, there will be a spouting
well — the water or oil, or both together, being
thrown out of the mouth of the boring. When
the tension of the gas is exhausted, resort must
be had to pumping, until the cavity is pumped
out. But in some cases a series of cavities com-
municate by small openings or crevices, in which
case a well may flow intermittently for a long
time, as it is replenished by percolation through
these channels. It is not uncommon for intermittent wells to throw out at first 300 or 400 barrels
a day, or to yield, in all, 20,000 bbls. They sometimes run two or three years before exhaustion.
When there is little or no gas, or where, from
the gas chamber being tapped, the gas is lost,
pumping has to be resorted to from the first.
Oil wells commonly vary in depth from 100 to
800 feet. Oil coming to the surface in springs is
not a reliable sign of oil cavities in the immediate neighborhood, for it is often carried a long
distance by the current of the subterranean
streamlets by which the springs are fed.

The oil of different districts varies considerably in specific gravity, and consequently in
value. The lighter oils are more valuable for
the purpose of illumination, and the heavier for
lubricators. The Oil Creek petroleum is usually
about 46° by Baume's hydrometer, being the
lightest oil found. At some of the wells it in-

creases in density to 38°. At Tidioute, on the upper Allegheny, the Economite well oil ranges about 43°. At Franklin the range is from 33° to 36°, and on French Creek and Sugar Creek the oil is also heavy, and is valuable as a lubricator. The West Virginia oil averages about 38°. The heaviest oil is found at Mecca, Ohio, the density being 26° to 27°, and the oil so thick that it will not flow in very cold weather. It bears a high price from its value as a lubricator. The oil obtained at Liverpool, Ohio, is of a similar character. The heavy oils are usually found in comparatively shallow wells, ranging from seventy to one hundred and eighty feet, whilst the lighter are commonly found several hundred feet below. The "third sand rock" of the Venango county system, in which the largest deposits of light oil are found, lies at a depth ranging from three hundred to twelve hundred feet. The majority of productive wells that have reached the third sand rock, range from four hundred to six hundred feet deep.

The yield of wells producing heavy, or lubricating oil, is generally much less than the average of successful wells of lighter oil, but, on the other hand, the value of the oil is much greater. A five barrel well of Mecca oil is equivalent in value to at least a twenty barrel well on Oil Creek. This fact must be borne in mind when comparing the value of wells in different localities. The "flowing wells" of large capacity run the lighter grades of oil, the heavy oils requiring to be pumped.

With regard to the condition of surface beneath which oil is most likely to be found in paying quantities, there is as much difference of opinion as there is in relation to the formation of the oil. In some places on Oil Creek, for instance, wells sunk on the flat bottom land are most productive, and those sunk in the side hill, or near it, find but little oil, whilst on the next tract the reverse of this becomes the rule. In and near Cherry Run several wells have been sunk far up the steep bluffs, and have proved successful. There appears to be no rule in the matter without a large number of exceptions.

One of the most tenable theories in regard to the production and distribution of petroleum in what now constitutes the principal field of its production in America, is that it is produced from the black or bituminous shale underlying the sandstone which is next beneath the coal measures. This bituminous shale extends considerably beyond the coal fields, cropping out on the shore of Lake Erie from near Sandusky Bay into the State of New York. Having less dip there than further towards the interior of the basin, it passes under the lake and into Canada.

Its western limit is a line passing from Lake Erie near Sandusky, Ohio, through Monroeville to Columbus and thence to the Ohio River, which it crosses between Portsmouth and Maysville. Exposure to the atmosphere robs the shale of some of its peculiar characteristics, the oil evaporating and leaving the stone of less weight and a lighter color; but by digging to a little depth it will be found of a brownish black color, emitting an odor of bitumen when rubbed or slightly heated, and will flame when placed in the fire.

From this bituminous shale the oil is distilled by the heat of the earth, and forced upwards into the substance of the sandstone, or through the cracks and fissures, by the pressure of the gas evolved in the process of distillation. The deeper into the earth the stratum of shale descends, the greater the heat, and the more complete the distillation. Hence, the oil found exuding from the rock, or obtained by shallow borings along or near the line of outcrop, is of a heavy character, frequently resembling tar, whilst the deep wells of Pennsylvania and West Virginia yield a lighter oil. It follows, as a matter of course, that the nearer to the centre of the basin the greater the depth of the deposit; the heat is more intense, the distillation more thorough, and the yield of oil greater. It does

2* B

not follow that the wells will have to be sunk
to a corresponding depth, for the expansive
force of the gas increases with the increased
volume of oil, and forces it up whenever a
fissure or disruption of the strata affords a
passage.

THE HISTORY OF PETROLEUM ON OIL CREEK.

THE existence of oil in the valley of Oil Creek, in Venango County, Pennsylvania, has been known for a long period. The Indians, from time immemorial, resorted to the valley at stated seasons to gather the oil for medical purposes; and the work of procuring it was prefaced and concluded with dances and other ceremonies, the final proceeding being to set on fire the surface of the pools covered with oil, and dance around the flame. There are evidences of the probable use of the oil by a race anterior to the Indians of our own period, and probably cotemporary with the mound builders of Ohio and the ancient miners of Lake Superior. In several parts of the Oil Creek valley, the early settlers found pits eighteen to twenty feet in depth, and from six to eight feet in diameter, carefully walled around with timber which the petrolized waters had preserved from decay, and in which

(19)

were found notched logs, which served as lad-
ders.   The Indians could give no account of
these pits other than that they must have been
dug by an earlier race, of a superior civilization,
frequently alluded to in Indian traditions.   Oil
is still found in those pits, and from them, and
from the surface of shallow pools, the Indians
obtained their chief supply.   It also bubbled up
in mid stream in many places, and was obtained
by throwing a blanket on the water, and, after
it became saturated, squeezing the oil into the
vessels prepared to receive it.   The early settlers
also used it as a medicine in cases of rheumatism,
and it was frequently sold in druggists' shops
for the same purpose, under the name of " Seneca
Oil."   An article in the "Massachusetts Maga-
zine" for July 1791, describes the oil springs in
what was even then known as Oil Creek, and
says that the American troops, in their marching
that way, halted at the spring, collected the oil,
and bathed their joints with it.   This gave them
great relief, and freed them immediately from
the rheumatic complaints with which many of
them were affected.   The troops also drank freely
of the waters, which operated as a gentle purge.

In the year 1845, Mr. Lewis Peterson, Sen., of
Tarentum, Alleghany County, Pa., brought to
the Hope Cotton Factory, at Pittsburgh, a sample, in a bottle, of what is now known as petro-

leum. It came up with the salt water from his salt well at Tarentum, and gave him considerable trouble. Mr. Morrison Foster, then of Pittsburgh, but now of Cleveland, in conjunction with the manager of the spinning department of the mill, Mr. David Anderson, experimented with the oil, and soon found that by a certain process it could be combined with sperm oil, in such a way as to form a better lubricator for the finest cotton spindles than the best sperm oil, which alone could previously be used for that purpose. The mixture cost about seventy cents per gallon, whilst the sperm oil alone cost one dollar and thirty cents. The saving was so great, in one of the heavy items of expense in a large cotton factory, that a contract was entered into with Mr. Peterson, by which the latter was to supply two barrels per week, and for ten years this oil continued to be used at the Hope Cotton Factory, unknown to any but the proprietors.

This is believed to have been the first practical use to which Petroleum was put in America. A few years afterwards Mr. Kier, who also had saltwells at Tarentum, and was troubled, like Mr. Peterson, by the oil that came up with the water, sent some of the oil to Prof. Booth, of Philadelphia, for analyzation. Acting on the advice of Prof. Booth, Mr. Kier took some of the oil to New York and experimented with it as a solvent for gutta percha. Failing in this, he was induced by Prof. Booth to try its merits as an illuminator, and succeeded in refining it so that it was used as "Carbon Oil" in Pittsburgh from 1850 to 1855, meeting with a sale that required all the oil to be obtained from the salt-wells of Tarentum to supply the demand.

About twelve years ago some attention was directed in different parts of the world to the subject of petroleum, or rock oil, and search was made for it in various directions. Among other places, Oil Creek became the object of attention, and a company was formed to procure oil from an oil spring, the existence of which had become known to a large number of persons. This company, which was organized by Messrs. Eveleth and Bissell, of New York, was known as the Pennsylvania Rock Oil Company, Prof. Silliman being at its head. Their operations were confined to collecting the surface oil, until, in 1858, Col. E. L. Drake, of New Haven, Connecticut, was engaged to visit the valley, and set about sinking a well on Watson's Flats, about a mile and a half below Titusville. The first well was unsuccessful, and another was sunk. This was a success. The drill struck an oil cavity at a depth of seventy-one feet, and, on the tools being withdrawn, the oil rose to within five inches of the surface. It was pumped off, and yielded at first four hundred, and afterwards a thousand gallons of oil per day.

As may be imagined, the excitement in the valley was very great. Every one that held land in the vicinity of the Drake well made preparations for sinking wells on his own account, or leased to others a right to sink wells, reserving to himself a royalty of from one-eighth to one-quarter the oil. Derricks were hastily put up, and "spring-poles" fixed, all of the early wells being sunk by hand. Some of the wells were successful, but by far the larger portion obtained no oil at all, or in such small quantities as to be unremunerative. The demand was small, the use to which the oil was put being as yet very limited. Still, several of the adventurers were making fair wages, when the discovery of flowing wells revolutionized matters. Pumping oil at the rate of five to twenty barrels a day was a discouraging process when, at another well, the oil was running spontaneously as many hundreds as the others were pumping single barrels. The glut of the market, caused by the flowing wells, and the consequent depression in prices, rendered the continuance in operation of the pumping wells a losing speculation, and nearly all of them were abandoned. The lessees fled in despair, in many instances leaving their machinery behind them, and not stopping to surrender their leases. Some of the abandoned wells have since been successfully worked, and more would be, but from the impossibility of getting at the holders of the old leases, and the fear to commence operations lest, at an unseasonable moment, the lessees should return.

The first flowing well ever struck was on the McElhinney or Funk Farm, and was known as the Funk Well. Funk was a poor man when the well was sunk. It was struck June, 1861, and commenced flowing, to the astonishment of all the oil borers in the neighborhood, at the rate of two hundred and fifty barrels a day. Such a prodigal supply of grease upset all calculations, but it was confidently predicted that the supply would soon cease. It was an "Oil Creek humbug," and those who had no direct interest in the well looked day after day to see the stream stop. But, like the old woman who sat down by the river side to let the water run itself out that she might cross dryshod, they waited in vain. The oil continued flowing with but little variation for fifteen months, and then stopped, but not before Funk became a very rich man.

But, long before the Funk had given out, the wonder in regard to it was overshadowed by a new sensation. Down on the Tarr Farm the Phillips Well burst forth with a stream of two thousand barrels daily. Not to be out-done by the territory down the Creek, the McElhinney tract "saw" the Tarr Farm, and "went it a thousand better." The Empire Well, close to the Funk, suddenly burst forth with its three thousand barrels daily, a figure subsequent flowing wells vainly endeavored to equal.

The owners were bewildered. It was truly "too much of a good thing." The real value of petroleum had not yet been discovered, and the market for it was limited. Foreigners would have nothing to do with the nasty, greasy, combustible thing. Our own people were divided in opinion. Some thought it a dangerous thing, to be handled at arm's length, whilst others set it down as a humbug in some way or other, of which the community should keep as shy as possible. The supply was already in advance of the demand, but the addition of three thousand barrels a day was monstrous and not to be endured. The price fell to twenty cents a barrel, then to fifteen, then to ten. Coopers would sell barrels for cash only, and refused to take their pay in oil or in drafts against oil shipments. Finally, it was impossible to obtain barrels on any terms, for all the coopers in the surrounding country could not make barrels as fast as the Empire could fill them. The owners were in despair and tried to choke off their confounded well, but it would not be choked off. Then they

3

built a dam around it and covered the soil with grease, but the oil refused to be dammed, and rushed into the stream, making Oil Creek literally worthy its name. For nearly a year it flowed, and then dropped to a pumping well, yielding about a hundred barrels. Lately it stopped, but on the application of an air pump, it revived, and is now steadily increasing its product, producing about a hundred and twenty barrels.

The Sherman Well, which was the next great "flowing well," was put down in the spring of 1862. It was sunk under great difficulties. J. W. Sherman, who was the original owner, commenced sinking it on the Foster Farm, next above the McElhinney, with very limited means, his wife furnishing the money. After a while it became necessary to procure an engine, but there was no money to make the purchase, and two men, who were in possession of the desired article, were admitted to a share for the engine. Soon after, when but a few feet more were necessary to reach the supposed deposit of oil, the funds were exhausted. A sixteenth interest was offered for $100, and refused. Ultimately it was sold for $60 and an old shot-gun. A horse became necessary during the work, and a share was disposed of for the animal. At last, when all the means that could be raised by borrowing or selling were about exhausted, oil was struck, and flowed at the rate of fifteen hundred barrels a day. The flow continued at this rate for several months, when it declined to seven hundred barrels. For twenty-three months the well continued flowing, and then it stopped. For the first year the proprietors made but little, if anything, owing to the low price of oil and the difficulty of getting it to market, but, during the second year, the market improved, and an immense fortune was realized. The well now pumps from thirty to forty barrels daily.

On the East side of the Creek from the Foster Farm is the Farrell Farm. Farrell was a poor man, employed in hauling oil, and was offered one-eighth interest in the land for $200. In March, 1863, the Caldwell well was struck on that farm, not far from the Sherman well, and flowed twelve hundred barrels daily. Two months afterwards, the well now known as the Noble and Delamater, but then as the Farrell well, close to the Caldwell, struck oil, and commenced flowing at the rate of two thousand barrels daily. The column of oil spouted up fifty feet, with a roar like that of a hurricane. For some days the oil ran to waste, there being no possibility of controlling its flow. As soon, however as its first fury was spent, a stop-cock was put on, and the flow reduced to a stream of the dimensions of a two and a half inch tube.

In the early days of oil enterprise, and after the yield had become large, considerable difficulties existed in getting the oil to a market. There were no railways to carry it off, and the only plan was to float it down the creek to the Alleghany, and ship it thence by steamer or flat-boat to Pittsburg. When the Atlantic and Great Western Railway was built to Meadville, a large number of barrels were hauled across the country by teams to that place, and shipped thence to New York.

The supply of flat-boats on the creek and river was far too small for the requirements of the oil trade. When boats could not be had the oil barrels were formed into a raft and lashed together. In this way they were floated, or towed, down to the mouth of the creek, where they were either loaded on steamers or towed to Pittsburgh. At times the great need was barrels. When this was the case the flat-boats were made oil-tight, and the oil poured into them in bulk. When there was not sufficient water in the creek, a large dam was made, and at an appointed time a pond freshet swept boats and rafts down to the river. Sometimes amusing, but expensive casualties resulted from these pond freshets. An unskilful boatman occasionally got his boat in the wrong position, and the whole mass of boats, rafts, and floating tanks was thrown into confusion. Rafts were broken up, tank-boats stove in, and an immense amount of property destroyed.

A far more dreadful disaster frequently happened. With the oil from flowing wells a large amount of highly inflammable gas escapes. The utmost precaution is generally taken to prevent any fire being brought into contact with the gas, but accidents are sometimes unavoidable. The first rush of gas, on a flowing well being struck, occasionally enters the engine shed, and takes fire from the furnace. A terrific explosion follows, and everything in the vicinity is wrapped in flames. When a well takes fire in this way it is very difficult to extinguish it. Water appears to have no effect, the only effectual way to extinguish the flames being to turn on steam, or stop the well hole by throwing dirt on it — a rather difficult task to perform in the near presence of an intensely hot "pillar of fire." Several of the leading wells have been on fire, and much damage done. Several times the boats on the creek took fire, and, breaking from their moorings, swept down stream, carrying devastation with them. A terrible fire of this kind occurred May 12th, 1863, when a great number of boats, loaded with oil in bulk and barrels, were on fire, and endangered the existence of Oil City. They

3*

swept down the Alleghany, destroying every
thing with which they came in contact. The
bridge across the Alleghany at Franklin was
totally consumed. The construction of railways
to the oil district, from Corry and Meadville, by
lessening the necessity for boats and rafts, has
greatly diminished the risks by freshets and fires.

In briefly sketching the history of a few of the
flowing wells, only those of the earlier and more
famous have been selected. A number of other
flowing wells have made their possessors wealthy,
and some have attained considerable notoriety.

The change in the fortunes of the original
owners of property in the oil regions must be a
source of wonder to themselves, as it is to every
one else. The so-called "farms" on Oil Creek
never produced enough to give decent support to
those who lived on them. The residents on the
creek led a rough life, generally eking out their
livelihood by rafting lumber to Pittsburgh, and
bringing from that city such articles as they
needed. So poor were many of them that they
were compelled to foot it home from Pittsburgh,
for want of means to pay for a conveyance. The
revenue derived from oil leases on their lands,
and fortunate speculations in oil and territory,
have made all of them wealthy, many of them
millionaires. Land that six years since was not

worth ten dollars an acre, has in some instances brought as many thousands.

Until recently, the wells in the Pennsylvania oil regions were owned by single adventurers, or by a few men associated together. The disadvantages in this mode of working were many and great. An adventurer who found his money and labor expended in the production of a "dry hole," rarely possessed means or perseverance enough to sink another well in the vicinity The well was abandoned, and gave a bad reputation to the whole neighborhood. With the present system of joint-stock companies, able to prosecute their work in spite of two or three failures, old property is more thoroughly developed, and the merits of new oil-bearing territory properly tested. The individual profits are not always so great, nor are the individual failures so ruinous. Oil mining becomes less a game of chance, and takes its place among those branches of business that offer a good prospect of profitable returns for the investment made in them.

The growth of the petroleum business is indicated in some degree by the following summary of exports in 1862–3–4. The exports in 1861 were small, and no accurate account was kept of them. It must be borne in mind that the con-

sumption in the United States is very large and
rapidly increasing; in fact, petroleum has become almost a necessary of life with us.

TOTAL EXPORTS IN 1864, 1863, AND 1862.

	1864. Gallons.	1863. Gallons.	1862. Gallons.
From New York	21,335,784	19,547,604	6,720,273
Boston	1,696,307	2,049,431	1,071,375
Philadelphia	7,760,148	5,395,738	2,800,973
Baltimore	929,971	915,866	174,830
Portland	70,762	342,082	120,250
Total exports from the United States	31,792,972	28,250,721	10,887,701

There were also exported, in 1864 from Cleveland direct to Liverpool 80,000 gallons refined.

AVERAGE PRICES FOR 1864 AND 1863.

	Crude.	Refined, free.	Refined in bond.	Naphtha refined.
Average 1864	41.81	74.61	65.03	39.54
Average 1863	28.13	51.74	44.15	28.53

HOW OIL WELLS ARE BORED AND WORKED.

THE individual or company intending to bore for oil, either purchases the land in fee simple, or obtains an "oil lease." At the present time the purchase of land in fee simple is mostly effected by companies, the high price put on oil-bearing lands rendering it almost impossible for individuals to obtain a tract of any considerable dimensions.

Of the tract thus purchased but a small proportion, generally, consists of what is now considered "borable territory," namely, the flat land bordering on the river or creek, and sides of a ravine, or bank of a stream. The remainder is usually high bluffs, valuable in proportion to the amount of wood obtainable for fuel. Some companies work their own property, whilst others grant oil leases to individuals or other companies.

An "oil lease" grants to the lessee a right to bore within certain limits for "oil, salt, or other

C   (33)

minerals," the work to be commenced within a
given time, and "to be prosecuted with all reasonable diligence." If these conditions are not com-
plied with, the land reverts to the owner of the
fee simple. Some of the leases granted in 1860
and 1861 were loosely drawn up, and made no
provision for the reversion of the property in
case of the abandonment of the work, and it
is no unfrequent thing, after an old well has been
taken by new adventurers and made successful,
for the original lessee to make his appearance
and claim compensation. If the new proprietors
do not comply, an injunction is obtained, and,
rather than have the work stopped for months
until the case comes up for trial, the victims are
generally willing to compromise. In the beginning of the oil enterprises on Oil Creek, as now
in some of the new oil territory, the owner of
the fee obtained, as compensation for granting
the lease, a royalty, or "landed interest," of one-
sixth or one-fourth the oil raised, leaving the
remainder, or "working interest," to bear all the
expenses. At the present time, on Oil Creek,
the landed interest obtains one-half the oil on
all new leases, and in some very desirable locations a bonus is demanded in addition. It will
be readily seen that the owner of the "landed
interest" gets the lion's share, receiving half the
product without being at any expense for work-

ing. The royalty was formerly paid in kind, but is now usually settled by taking half the value of the sale of oil. Should the lessee abandon his adventure he is allowed to remove his derrick and engine.

Having bought or leased a location, the next step is to select the exact point for boring. In this the experienced worker is guided, to some extent, by the nature of the soil and the position of the ground. A new class of people has sprung into existence under the cognomen of "oil smellers," who profess to be able to ascertain the proper spot for boring by smelling the earth. Some of them practise considerable mummery in order to mystify and impress their employers. The "witch-hazel" is also frequently used, the professional locater of wells marching solemnly along, holding his hands apart with one end of a forked hazel-rod in each. On passing over an oil spring or basin, the point at the junction of the forks suddenly deflects towards the earth, and there the work is commenced. As the witch hazel has the property — according to believers in its powers — of finding streams of water in the same manner, it sometimes happens that water, instead of oil, proves to be the product of the well.

The exact spot being determined, a huge derrick is erected immediately over it. This is a square frame of timbers, substantially bolted together, making an enclosure about forty feet high, and about ten feet at the base, tapering somewhat as it ascends. This is generally boarded up a portion of the distance to shelter the workmen. A grooved wheel or pulley hangs at the top, and a windlass and crank are at the base. A short distance from the derrick a small steam engine, either stationary or portable, is fixed, and covered with a rough board shanty; a pitman rod connects the crank of the engine with one end of a large wooden walking-beam, placed midway between the engine and the derrick, the beam being pivoted on its centre about twelve feet from the ground. The walking-beam is a rude imitation of that of a side-wheel steamer. A rope attached to its other end passes over the pulley at the top of the derrick, and terminates immediately over the intended hole. A cast-iron pipe, from $4\frac{1}{2}$ to 5 inches in diameter, is driven into the surface ground, length following length until the rock is reached. In the older wells the ground was dug out to the rock, and a wooden tube put in it. The earth having been removed from the interior of the pipe the actual process of boring or drilling is commenced. Two huge links of iron, called "jars," are attached to the end of the rope. At the end of the lower link a long and heavy iron pipe is fixed, and in

the end of this is screwed the drill, about three
inches in diameter, and a yard long. When all
is ready the drill and its heavy attachments are
lowered into the tube and the engine set in motion. With every elevation of the derrick end
of the walking-beam, the drill strikes the rock,
the heavy links of the "jars" sliding into each
other and thus preventing a jerking strain on
the rope. The rock, as it is pounded, mixes in a
pulverized condition with the water constantly
dripping into the hole, and assumes a pasty form.
After a while the drill is hoisted out and a sand-
pump dropped into the hole. The sand-pump is
a copper tube, about five feet long, and a little
smaller than the drill, having a valve in its bottom opening upwards and inwards. As the tube
is dropped into the hole the pasty mass rushes
into it through the valve and remains there.
When this has been done several times the tube
is hoisted out and emptied, the operation being
repeated until the hole is clear, when the work
of drilling recommences. It is evident that as
the drill is not round at the point, but with a
chisel-shaped edge, the hole would not be round
unless some other means were adopted. This is
partially accomplished by the borer, who sits on
a seat about six or eight feet above the hole, and
holds a handle fixed to the rope, giving the latter
a half twist at every blow. By this means a

4

nearer approach to a cylindrical hole is attained. But the hole must be as nearly round as possible, and therefore the tools are taken out, and a "rimmer," or "reamer," sent down, which cuts down the irregularities of the hole.

In the earlier days of well-boring, as now in some localities, the wells were sunk by hand, or by horse-power. In the former case a stiff spring-pole, firmly secured at one end, lifted the drill and rods suspended from its free end, and the power was applied to this end to make it suddenly descend. Two men, standing together, placed each a foot in a double stirrup suspended from the pole, and suddenly bore it down. Immediately it sprang up, and the operation was repeated. This was a tedious and laborious operation, and has been generally abandoned.

As the holes get down to points where the first indications of oil are reached, the contents of the sand-pumps are anxiously examined. The oil-borers have a geological system of their own, the prominent points of which are three layers of sandstone. The "first sandstone" lies immediately below the alluvial deposit. The "second sandstone" is at a variable depth of 100 to 300 feet, and here the first indications of oil are reached. Some wells go no lower than the second sandstone, but the general plan is to go down into the "third sandstone," where the largest and most reliable deposit of oil is found.

It frequently happens that the drill breaks and falls off, and becomes fixed in the hole. Nothing can be done until the tool is removed. The remaining portion of the boring instrument is taken off, and a pair of nippers or clamps let down into the hole to grip the broken drill and extract it. Some men make the extraction of tools a special business, and exhibit great ingenuity in their devices to overcome the difficulties they have to encounter. There are instances where wells have had to be abandoned in consequence of the tools remaining immovably fixed in the hole.

When the hole has been sunk to a sufficient depth and "strike oil," the next thing is to extract the oil from the well. If a flowing well has been struck, all trouble on this head is saved, as the oil and gas rush out in a stream, sometimes with such violence that the men have to make their arrangements with considerable rapidity, or the precious fluid runs to waste. The first business is to tube the well. An iron pipe, with a valve at the bottom like the lower valve of a pump, is run down the entire depth of the well, the necessary length being obtained by screwing the sections firmly together. If the oil does not flow spontaneously, a pump-box, attached to a wooden rod, also made of sections screwed into each other, is inserted in the tube, and the upper end of the rod attached to the "walking-beam." The well is now ready for pumping.

One important feature in the tubing process must not be forgotten. In boring for oil, springs of water are of course cut through and the water falls into the hole. Being heavier than the oil, it lies at the bottom, and would enter the pump-tube but for a very ingenious contrivance known as the seed-bag. This is a leather bag, in shape something like a boot-leg, filled with flax-seed, which is fastened around the iron tube at what is considered the proper point, and crowded down with it. When the seed-bag becomes wet it swells and thus forms a water-tight packing between the tube and the rock. At times the seed-bag slips or bursts, the well at once fills with water, and the tubing has to be pulled in order to refix the seed-bag.

More or less gas accompanies the oil in its passage to the surface. If a flowing well, the gas is allowed to escape, there being no use for it, and it can be distinctly seen puffing out of the pipe, generally with labored breathings or panting, the cause of which is known among the operators as the "breathings of the earth," in reality being the irregular obstructions to its passage by the unequal flow of oil in the bottom of the hole. The passage of the oil from a large flowing well is a curious and interesting sight.

In many of the pumping wells the gas is saved and used, either by itself or with coal, as fuel for the engine. To save it, the mingled gas, oil and water — for in spite of all precautions some water will come up from nearly every pumping well — is conducted by a pipe from the well tube into a tight barrel. The oil and water fall into the bottom of the barrel, and run off by a pipe near the bottom into a huge tank or vat, where another separation is caused by the different gravities of the two fluids, the water sinking to the bottom of the vat. The gas escapes by a small pipe at the top of the barrel, and is conducted into the furnace, where it burns with a fierce and steady flame. The engine of the Forest City well, as also many other wells, is run entirely by gas, the jet being spread into a broad and waving flame by passing through a piece of sheet iron pierced with holes. Its steadiness is shown by the fact that the engine house is lit with several jets of gas, of a steadier and purer flame than that furnished by some gas companies.

The oil, as it flows into the tank, is a dark green fluid. When sold for shipment it is drawn off by a faucet in the bottom into barrels. In the larger wells, where a considerable quantity

4*

of oil is kept on hand before sale, ranges of vats are built, the oil flowing from one to the other. The vats are covered with boards, and at the larger wells roofed in to prevent evaporation. At the gassy wells great care has to be taken with regard to fire, as a lighted cigar might set fire to the gas and blow up the whole concern. In the early days of the flowing wells, before their nature was thoroughly known, serious con- flagrations took place from this cause. Should a well take fire, water not only fails to extinguish it, but seems to add to the fury of the flame.

Experiments are in progress for testing the practicability of a new process of boring. The principle is that of cutting out a hole instead of pounding it. The drill is circular and hollow, being a thin tube, set at its lower edge with Brazilian diamonds, of hardness sufficient to cut glass. It is connected by an iron rod to bevelled cog wheels attached by cranks and rods to the walking-beam of the engine. The surface of the upper rock being cleared, the drill sits on it and revolves with great rapidity, cutting its way down at a rate astonishing to old well borers, and leaving a central core standing. A clamp is let down which grips the core and jerks it up in the form of a perfectly smooth cylinder. Water is poured down the hole to assist the cutting pro- cess, until the natural flow from the springs cut supplies the want. The portions of the core shown exhibited the stratification of the rock, and will go far to settle some vexed questions about the strata which cannot be ascertained by the ordinary method of drilling.

Five feet of rock had been cut at the rate of four inches in five minutes, or ninety-six feet per day, when some changes were required in the machine, and it was removed for alteration. The patentee is satisfied that he can put down a well five hundred feet in ten days, at no greater cost to the well-owner than by the present tedious process, which takes from two to four months. Another boring machine, of somewhat similar design, has also been introduced, but neither has yet been thoroughly tested.

It sometimes happens that after a well has been yielding for months it stops and refuses to yield another drop. This is occasioned in some instances by the thickening of the paraffine at the bottom of the hole, and the consequent obstruction to the flow of the oil into the pump box. To remedy this a jet of steam from the engine is introduced and forced down the hole, melting the coagulated mass, and restoring the flow into the pump. Another plan, which is coming into use, and which has so far proved successful, is to use an "ejector," or air-pump, with two pipes inserted into the tube of the well.

The air is forced down one pipe into the vein at
the bottom, and the oil rushes up in a steady
stream through the other. By the use of these
" ejectors " a number of wells have been restored
to a yield ranging from thirty to a hundred and
forty barrels daily, after they had been considered
worthless by their owners.

An invention has recently been introduced, and
has proved highly successful in the experimental
tests to which it has been subjected. This is
known as the "Roberts torpedo," and consists of
a cylindrical tube four feet in length, made to fit
the bore of the oil well, which is filled with gun-
powder, lowered by a wire into the well to any
desired point, and exploded by the percussion of
an iron weight or follower, sent down the wire
after the torpedo is lowered. The torpedo, on
being exploded, drives out the paraffine or other
coagulated matter, and at the same time opens
fissures into the solid rock, very often leading
into veins or new wells of oil. An experiment
on this principle was made at Marysville, Medina
County, Ohio, in 1861; but from the imperfect
nature of the means used, it failed as a practical
measure, although the correctness of the theory
was proved. The rock was shattered and oil
forced up in various places at a distance of several
rods, but the well was ruined by the iron pipe
used being jammed into the hole.

The cost of sinking a well differs greatly according to the locality, the nature of the rock, the depth to be sunk, and other incidental causes. The engines used range from six to eighteen horse-power, the general size now coming into use being from ten to twelve horse-power. The mistake frequently made by inexperienced managers is the getting of too small engines, the object being the saving of expense. It is a false economy. Small and inefficient engines are the cause of much of the trouble at the wells. It is better to err on the side of unnecessary power, as the same engine can be used for sinking new wells whilst pumping the original well. In some places five or six wells have been worked by the same engine. Get a good engine, in perfect order, and it will be found the most economical. Such an engine will cost from $2,100 to $3,000. The derrick, walking-beam, &c., will cost about $150; one set of drilling-tools, about $300. The inch and a half hawser, sand-pump, tubing, drawing-pipe, and miscellaneous tubes, will average $1000 more. The drilling is generally done by special contract, the owners of the property finding the tools and machinery. The average contract price is $2.50 per foot for the first five hundred feet. Where the proprietors determine to sink wells with their own workmen, it is necessary to keep a strict watch, as the "superintendant" is very apt to spend a considerable portion of his time in "prospecting" and land speculating on his own account, and the workmen, being paid for their tedious and monotonous work by the day, are usually in no hurry about it, when left to themselves. The total cost of sinking a well five hundred feet may be estimated at from six thousand to nine thousand dollars. The first estimate may not be exceeded, but so many mishaps occur from the breaking of tools, giving away of machinery, and other causes, that it is better to have a wide margin. The second well can be sunk at about half the expense, by using the same engine and tools.

After the well is sunk, should it flow, the expense is merely nominal. If it is a pumping well, two engineers, one or two extra hands, and the fuel, will make the ordinary daily expenses from ten to twenty dollars, according to circumstances. The breakage of the wooden pump-rods, giving out of machinery, delays by slipping and bursting of the seed-bag, and the frequent difficulty of obtaining fuel, even at a high cost, will frequently increase the expenses and diminish the receipts at the same time.

THE PENNSYLVANIA OIL REGION— THE PRINCIPAL OIL LOCALITIES AND HOW TO REACH THEM.

PENNSYLVANIA is, at present, the greatest oil-producing State in America, or the world, and Venango county is the principal oil region of Pennsylvania. Some developments of oil have been made in Crawford, Clarion, Fayette and other counties, but so far, Venango county has been the chosen seat of empire of King Petroleum. If a line be drawn nearly through the centre of the county, running from north to south, tending a little west, it will pass along Oil Creek, the central and most productive portion of the oil territory. From Franklin, a few miles below where Oil Creek joins the river, the Alleghany to the east, and French Creek to the west, form a huge V, with Oil Creek passing down the middle and joining the right arm of the V just above the point of junction. Below this point the Alleghany stretches, converting the V into a

Y. The centre, or Oil Creek line, is that of the greatest yield at present, the others not having been so extensively worked. The first discoveries of oil were made on Oil Creek, and for some time explorations were confined to that line. The success there met with induced others to examine into the oil-producing qualities of the adjoining streams, and a number of holes were sunk on the banks of the Alleghany River, French Creek, Sugar Creek, (an affluent of the French,) and some of the "runs," or small streams, tributary to the several creeks.

Beginning at the point where the Alleghany crosses the Venango county line, a short distance below Tidioute (the highest point of oil operations on the river), the oil line stretches along the banks of the Alleghany, in its sinuous course, for about sixty miles, during no part of which distance would a voyager be out of sight of the derrick of an oil well, past, present, or prospective. After entering the Venango county lines, the principal streams discharging into the Alleghany on its way south, are the East Hickory on the west; Tionesta and Little Tionesta, Hemlock Creek with its branch known as Porcupine Run, on the east; Culberton's Run and Pithole Creek on the west; Horse Creek on the east; Oil Creek, Two Mile Run, and French Creek on the west; East Sandy on the east; Big Sandy,

Big Scrub Grass and Little Scrub Grass on the west. Besides these there are numerous smaller streams that have not yet attained notice for their oil-bearing qualities, although many companies have been organized for their development. All the streams mentioned have become oil locations, and on each of them the work of pumping or boring is going on with great activity.

From the mouth of Oil Creek to Titusville, just across the Crawford county line, is a distance of twenty miles, along the whole of which the wells are thickly planted. Ascending the stream, Cornplanter Run heads off towards the north-west, Cherry Run to the north-east, Cherry Tree Run, and Weikel Run which branches from it, to the north-west; and Bennehoff's Run to the west. Just below Titusville, Oil Creek forks to the east and west. All the tributaries of Oil Creek are oil-producing, and are crowded with wells.

French Creek, one of the largest affluents of the Alleghany in Venango county, comes in from the north west. A number of wells are scattered along its banks. Sugar Creek enters French Creek from the east about three miles above Franklin, and is now a favorite oil locality.

Meadville is the central point of departure from which to reach any part of the Venango county oil regions. From it the traveller can

5   D

enter Oil Creek at either the Oil City or Titusville end. The best route is by way of Franklin
and Oil City. Arriving at Meadvile by the
Atlantic and Great Western Railroad, the visitor
can obtain a comfortable night's rest and an
excellent breakfast at the McHenry House, the
hotel in the depot building. Taking the Franklin Branch cars a little before eight o'clock in the
morning, the distance to Franklin, twenty-eight
miles, is done in something under two hours and
a half. The railway follows the course of French
Creek throughout, affording in summer a series
of picturesque scenes. The last five or six miles
of the route is lined with oil wells, nearly all put
down in 1860 and 1861, and abandoned. A few
have resumed work.

From Franklin to Cooperstown, on Sugar
Creek, is about eight miles over a fair road.
Conveyances can be had in Franklin. A nearer
route to Cooperstown is to leave the train at
Utica station, nineteen miles from Meadville, and
take the road across. This will save from two
to three miles, but the chances of obtaining conveyance across are not many, as there are no
livery stables in Utica.

Horses or conveyances of some kind can be
obtained in Franklin to visit the Alleghany
below. The visitor must not rely too much on
this, however, as the great rush of people to the

oil regions makes a greater demand for conveyances than can be met by the limited supply. As might be supposed under such circumstances, prices rule extravagantly high. Livery stable keepers charge about ten per cent. on the value of a horse when letting it out for a day. The Alleghany below Franklin is very crooked, and the distance by the river bank is much greater than by the roads that keep a short distance from the stream.

Until recently, the communication between Franklin and Oil City, seven miles, was by steamer twice a day in summer, and by private conveyance or foot in winter. The Franklin Branch Railway is now open for passengers to Oil City.

From Oil City up the Alleghany, there is a road hugging the river bank, and crossing the river by ferries at several points where the steep bluffs block the way. To reach Pithole Creek, and the river above that point, the best route is to go from Oil City to Plumer, seven miles, and turning to the right from the centre of the village, cross Pithole Creek, and strike the river at Culbertson's Run. From this point a good road extends along either bank. Two villages are passed on the road before reaching the northern line of Venango county. At President, near the mouth of Hemlock Creek, a large hotel has recently been built. At the mouth of Tionesta Creek is the village of Tionesta. Both these villages are on the east side of the river. From Oil City to Hickory Creek is about twenty-eight miles.

From Oil City up Oil Creek to Titusville the choice lies between horseback and foot travel. The best way, on every account, is to walk. The horses betray few traces of Arabian blood, and their habits are too devotional for comfort or safety. The greater number drop on their knees at every opportunity. By going on foot the visitor can see more, and, in many instances, travel faster than on horseback. From Oil City to Shaeffer's Farm, where the Oil Creek Railroad is first reached, is about twelve and a half miles of about the worst road — or rather no road — in the United States. There are several stopping places on the route, Rouseville, McClintockville, and Petroleum Centre having tavern accommodations—such as they are. At Shaeffer's Farm the train can be taken in the evening for Titusville, seven miles, and, early next morning, from Titusville to Corry, twenty-eight miles, and back to Meadville, forty-two miles farther.

Everywhere the visitor must be prepared for rough living and hard lodging, if fortunate enough to obtain lodging. When intending to stop at night at any particular place, telegraph in the morning to engage a bed. By doing so you will have a slight chance of obtaining half a bed. If this is neglected there is a certainty of getting no bed at all.

Wear such clothing as will excite no regrets should they be covered with mud or grease, as they inevitably will. Put on long legged boots, made waterproof. Carry no baggage except a small travelling bag or haversack, suspended by a strap over the shoulder. A blanket will be found very convenient in case no bed can be obtained, or as an addition to the scanty amount of bed-clothing, should a bed be secured. A lunch, or some crackers and cheese, in the haversack will be found convenient in case a tavern cannot be reached by dinner time.

The visitor to the Oil Regions who cannot "rough it," amid mud, filth, grease, wretched roads, deep quagmires, miserable accommodations, and poor food, had better stop at Meadville, eat a hearty dinner at the McHenry House, and then take the first train for home. He has not had a "call" for life among the oil wells.

5*

A TRIP DOWN OIL CREEK — MEADVILLE TO SHAEFFER'S FARM, VIA CORRY.

THE narrative of the leading features of a three weeks trip through the oil regions of Venango county, during which all the important localities were visited and thoroughly explored, will give some idea of the nature of the country and the business done in it, but no description can do the subject proper justice. An actual visit can alone give one a proper appreciation of the vast importance of the petroleum business.

Travelling in the roomy and elegant cars of the Atlantic and Great Western Railway, the journey was performed with comparative ease and comfort. At Meadville we halted for the night at the McHenry House, that we might enter the oily land with daylight to reveal its wonders. Here we found the principal topic of discussion was oil. The wave of excitement which was said to be sweeping through the valleys to the southward, rippled gently in the

(54)

McHenry House, and people were discussing the latest news from "the Creek." Every one we met with was "in oil," and every one was making arrangements to get deeper into the grease. Big stories were told of the fortunes made at the wells, and by the owners of oil lands, and bigger tales of the frightful state of the roads. I dreamed all night of thousand-barrel wells throwing up oceans of mud, and wading in greenbacks to the knees. My travelling companion in the opposite bed interrupted his "distinct breathings," with muttered offers of "ten thousand dollars for the refusal of your farm for five minutes," awakening me with his demands for an immediate answer. I set him down as a pitiable case of "oil on the brain," and tried to go to sleep.

At five o'clock of a dark, cold, and snowy morning, we set out by a freight train for Corry, having determined to enter the oil region by the Titusville route. That it is not the most convenient route was a fact of which we soon had abundant evidence, but, on the whole, there was not much to complain of, although travelling in a caboose car very early on a cold morning, is not the most pleasant experience in the world.

Everything must have an end, and shortly after eight o'clock we reached Corry. Here the Atlantic and Great Western, Philadelphia and

Erie, and Oil Creek Railways meet. The junction station is a miserable little affair, of rough boards, and utterly unable to shelter one-half the crowd waiting to go by the different trains. The snow was driving furiously, the weather was getting momentarily colder, and every one sought shelter. The dense mass, packed into the miserable little station like herrings in a cask, formed a motley assemblage. There were but few women among them. The men were of all ranks, ages, and descriptions. Sharp-eyed, trim-dressed, and eager speculators from New York, Philadelphia, and Pittsburgh, carpet-sack in hand, or with travelling bag strapped over the shoulder, going down to secure "a big thing;" traders anxious to open up a line of custom; rough fellows, going down to work at the wells; and old farmers, coarsely clad, and with their cowhide boots covered to the tops with mud whose layers spoke of months of travel over villanous roads, just as the concentric rings of bark on a tree reveal its years of growth, but who had within a year or two been made rich by farms that had previously made them poor — all were bawling for tickets for "Titusville" or "Shaeffer's Farm," until the ticket-clerk was well nigh driven desperate.

For nearly an hour the crowd surged outward towards the platform, as the rumble of a passing locomotive was heard, and inward towards the store, as the origin of the sound became known. Just as the crowd had settled down to the conviction that there was to be no conveyance to Oil Creek, a shout of " Train " was heard.

A bomb-shell suddenly dropped in their midst could not have produced a greater stampede than did that shout. The train was slowly backed down to the station, when the crowd rushed furiously at it. They swarmed up the steps, into the baggage car, over the locomotive, everywhere but under the wheels, and how they escaped that was a mystery. All the courtesies and amenities of life were disregarded. Men fought for precedence as if their lives depended on it. Women were rudely thrust back by anxious men who clung to the step-rails and kicked off those who endeavored to climb over them. Three cars and a baggage car were in three minutes packed almost to suffocation. A rattle and a jerk, and the train was off, shaking and jolting every one into position. We were well on our way to the " Oil Dorado."

From Corry to Titusville the railroad passes through an irregular country, and the track generally follows the original configuration of the land. Up hill the huge locomotive pants with its heavy load, and down hill it rushes, shrieking as if anxious to plunge itself into destruction.

Corry, with its scattered houses, its immense oil refinery, the largest establishment of the kind in existence, and all the other items that make up a thriving town, where three or four years ago there was nothing but "the forest primeval," is soon left behind. So also are the long trains of engines waiting to be united to innumerable derricks already lining the creek, but which will have to wait longer yet, owing to the inadequate facilities possessed by the road for transporting the immense amount of freight crowding on it. Soon the line of the creek is struck, and the road skirts its edge, most of the way winding along a ledge cut in the face of almost perpendicular cliffs. Here and there a derrick, like the skeleton of a church spire with its apex sawn off, and the frame not yet lifted on the church, keeps solemn watch along the banks, pickets of the advancing army of Petrolia. Presently the derricks increase; they close up their ranks, and soon stand in unbroken line along the left bank of the stream, throwing frequent skirmishers across to the right bank, effecting lodgment at the foot of the precipitous cliffs, where there is barely room to stand, and even threatening the railroad track which winds higher up. Puffs of steam and creaking of engines show where the pumping wells are at work. The river is dark, and a scum of oil glistens on its surface. Here and there a small board-shanty, connected by slender pipes with tanks at a little distance, marks the existence of a refinery — for all the processes connected with oil, from its extraction from the rock until it is ready for consumption, are carried on in the vicinity of the wells, employing a great number of refineries in addition to those in successful operation at Cleveland, Pittsburgh, New York, Philadelphia, and other places. The river margin widens and the number of derricks increase. No longer in single line, they double and treble their ranks, and appear in unbroken column; the new timber showing the large proportion just started, and the black and greasy appearance of many proving that their owners have " struck ile."

Twenty-eight miles from Corry the train stopped at Titusville, the last point in Crawford county before entering Venango county. A few years ago, Titusville was a lively little village, chiefly inhabited by lumbermen and raftsmen. In 1855 it was credited in the Gazetteer with having "an Universalist church and 243 inhabitants." Now it has a population of over six thousand, and is rapidly increasing. The one church has found several others to keep it company. There are thirteen hotels, crowded nightly with guests, of whom a large proportion have to spend the night without the privilege

of half a bed (an entire interest in a bed is a
thing unknown in the oil region). Two banks
do a large business in the funds produced by
operations in oil, and a third bank is nearly ready
to open. A new and handsome reading-room,
well supplied with the papers and periodicals
of the day, has been opened, and there is a hall
kept constantly engaged by lecturers, concerts,
or other popular amusements. In every part
Titusville gives evidence of its state of transition
from a small village to a thriving city. Lofty
and handsome brick blocks alternate with small
dilapidated wooden buildings. A well-made plank
sidewalk borders a muddy canal, by courtesy called
a street. When the citizens have time, some day,
they will probably rectify those little irregularities,
but just now every one is too busy. There are oil
wells within the limits of the town, and some of
the new settlers who have purchased lots on some
of the streets are in doubt whether to erect a
dwelling or a derrick. One is necessary, yet the
other may pay best.

The platform at Titusville station was crowded
with people, some waiting to see the new arrivals,
but most intending to take the train for farther
down the creek. For every person who left the
train at least three got on, so that the crowd became even thicker than before, and a number
were driven to the platform of the cars.

From Titusville to Shaeffer's Farm is seven miles, and all the way there were abundant evidences of oil adventurers, past, present and prospective. About a mile and a half below Titusville, on Watson's flats, is the scene of Col. Drake's first experiment in sinking oil wells, the result of which has been the enriching thousands of persons, and the addition of an immense business to the resources of the nation. Near this point comes in the East branch of Oil Creek, which has now been purchased and leased to nearly its entire length, for the purpose of boring for oil. Along the whole route to Shaeffer's Farm the derricks increased in number until there was a perfect forest of dismantled steeples. The air was redolent with the greasy perfume, and the passengers in the crowded cars talked more fiercely about oil, and discussed vast sums of money more glibly.

Miller's Farm, at which the train stops for a few minutes, is now a scene of busy activity. In the Autumn of 1864 three-fourths of the farm, comprising a tract of three hundred and seventy-five acres, was purchased by the "Indian Rock Oil Company," of New York. The enterprise was a vast one, for the purchase of so large a property in the very heart of the developed oil region, required large capital.

The property at the time of its purchase con-

6

tained a number of wells in operation, and others going down. The new proprietors, anxious to fully develope the value of their property, instead of floating their stock upon the market, proceeded at once to the work of sinking new wells, and in this way have already expended about $75,000. The result of this course will, from present appearances, be the production of several new successful wells.

Shaeffer's Farm at last. The crowd tumbled out of the cars as frantically as they clambered in, and, clutching their scanty baggage, rushed wildly for the "hotel," scrambling over each other in their anxiety to get first at the register. Every man as he scrawled his name with a nervous hand, inquired if he could get a room at night, and was met with the chilling response from an individual in high boots, covered with mud, that "there was almighty little show for anything, as it looked to him." Determined to make sure of what was at hand, and trust to luck, for the future, the crowd broke for the dining-room, not stopping to go through the ceremony of washing, for the land of grease and dirt has been reached, and the niceties of civilized life are henceforth disregarded. A plunge was made through the narrow passage to the dining-room; already keen-scented nostrils snuffed the titillating odor of roast and boiled, and hungry mouths watered in expectancy. But an impassable obstacle presented itself in the shape of a grim janitor who refused admittance without a ticket, and the "Johnny Newcomes" had to fight their way back to the bar and deposit seventy-five cents for the bit of blue paste-board, whilst the old stagers who were better provided, entered and filled all the vacant chairs. To give an idea of the rush of pilgrims to the Oily Land the fact may be cited that at one tavern at Shaeffer's Farm, about four hundred people dine daily. As to the quality of the meal we have nothing particular to say. The price was first class, and if the viands fell short of first class standard, the people at the "oil diggins" have no business with nice stomachs.

Dinner bolted, our first enquiry was about getting down the Creek. Conveyances there were none, from the fact that there were no roads to travel on. A single glance at the coun try around the "hotel" settled that question. A walk of three feet from the door in any direction brought the wayfarer into mud knee deep— and such mud! Clayey, slippery, greasy, sticky mud, into which you slid easily to uncertain depth, but which clung with fond affection to your legs, and endeavored to perform the offices of a boot-jack; deceptive mud, that appeared of uniform quality, but which in places suddenly engulfed the traveller thigh deep. Some of the pilgrims struck out boldly but were soon stuck fast, monuments of their own rashness. Clearly that mode of travel was not to be thought of except in case of dire necessity.

A good Samaritan appeared on the scene in the person of an exceedingly dirty and rowdyish looking young fellow, with the guise of a canaller. He loudly invited every one to take the packet boat for Oil City. Here was hope — doomed, alas! to be crushed as soon as born. The packet boat was an oily flat-boat, without shelter or seat, and the fare for the twelve miles by this precious conveyance was only three dollars and a half, or about thirty cents a mile! So that plan was rejected, and a brief council of war resulted in the decision to stay all night at Shaeffer's Farm, and start down on foot early in the morning. By dint of considerable finessing we secured a half-interest each in a small bed, packed with another bed in a dark closet dignified with the title of a "sleeping-apartment." As four persons occupied the room, in which there was barely space enough for two persons to undress at one time, and as there was not a window or opening of any kind for ventilation, but little clothing was required to keep us warm, a fact of which the landlord was evidently aware. Our neighbors "across the way" were deep in oil. and kept up a continued conversation on the subject. About two o'clock in the morning I dropped asleep, lulled by a confused sound of "flowing well — five hundred thousand dollars— one-half the oil—two years ago he wasn't worth a red cent—two thousand dollars a day—the biggest thing yet—third sandstone—made his everlasting fortune." My last mental reflection was that I wished I could say so of myself.

6*   E

DOWN OIL CREEK TO OIL CITY.

He who essays the "middle passage" between Shaeffer's Farm (the present terminus of the Oil Creek Railroad) and Oil City, must prepare himself for an experience for which life in the city affords but a poor preparation. The second step from the hotel at Shaeffer's plunges the pedestrian into a sea of mud which extends with varying depth to Oil City, more than twelve miles, with scarcely a friendly rock on which to rest the sole of the foot. Mud everywhere, illimitable, unfathomable. Let him who thinks he can make the passage by turning up his trowsers over his ankles and picking his way, at once disabuse himself of the idea. If he does not, ten steps from Person's Hotel at the Shaeffer will do it for him.

Lest any intending visitor to the oil regions should be discouraged by this picture and confine his wanderings to the limits of the railroads, I warn him that if he would see anything at all

of oildom he must make the passage, unpleasant
as it may be. There is no alternative. To see
the tips of the elephant's ears, or the end of his
tail, is not to see the animal, or form any idea of
his bulk, and there is no other way of doing it
than to "wade in." And this much may be
added, that whoever makes the trip, with his
eyes open, will never regret it. The sight is one
of which no description, however graphic or minute, can give a just idea.

The best way of making the passage, whether
in the muddy season, or in the season of ice, is
to travel on foot. It is the most independent,
enabling the visitor to pursue his investigations
with greater freedom, and is, moreover, in general, the most expeditous way. A flat-boat is an
abomination, and a horse — especially such as
they have on the Creek — is vanity and vexation
of spirit. Strike out boldly on foot, and pull
your legs up when they disappear from sight,
remembering that if you descend deep enough
you may strike oil. There is a choice of paths
in going down or up the Creek, the difference
between them being that each is muddier than
the other, and that you are certain to select the
muddiest.

The morning we set out from Shaeffer's, heading
down creek, was intensely cold, with some little
flying snow. The ground was frozen hard, with

ridges and knobs, making the travelling even worse than it would be in soft mud. Not unfrequently what appeared to be solid ground would prove to be a mere thin crust, covering a deep mud-hole, into which an unwary step would send the unlucky traveller knee-deep, sorely to the wear and tear of Christian patience and forbearance.

After leaving Shaeffer's Farm the route lay through the Stephenson and Gregg farms.

With every rock and turn of the sinuous creek the derricks increased in number, and the wheeze and clank of the engines grew louder and more confused. Climbing around the bluffs, over a steep path, then striking the newly graded track of the unfinished Oil Creek Railroad, chiselled out of the face of the cliffs, and at last descending to the half-frozen mud of the valley, we came out on the Foster Farm, crowded thickly with derricks and engines, groaning and creaking with the labor of pumping up the liquid treasures of the earth, more valuable than the golden waters of the ancients. About sixty derricks were massed together on this little tract of land, most of them with their black, greasy vats, sometimes ranged in a row, capable of holding each from five to ten thousand dollars' worth of oil, and several of them full. New derricks were going up, and engines stood around, waiting to

be put up in the proper places and set to work.
By the roadside is a row of vats, at one end of
the row being the time-stained and greasy derrick of the famous Sherman well, whose history
has already been given. Near it is the Porter
well, which in May, 1864, commenced flowing
about a hundred and fifty barrels daily, but now
pumps from fifteen to twenty barrels. On this
farm are the wells of the Briggs Oil Company,
and across the Creek are those of the Gillette
Oil Company, both under the management of
Mr. J. T. Briggs, who also manages the property
of the Indian Rock Oil Company. The Briggs
Company has been enabled to pay heavy dividends from its receipts. The Gillette Company's
wells, on the Espy Farm across the Creek, are
partly producing and partly boring, the prospects
being very good for the proprietors. Close to
these wells are the old Buckeye and the Buckeye
Belle wells. The Buckeye formerly flowed largely,
and bears an extensive reputation from the fact
that it was the oil of this well which Mr. J. T.
Briggs shipped to Europe as a sample, being the
first American rock oil ever sent across the Atlantic. Having lain idle for some time the well
became choked, but has been restored by an air-
pump and is now doing well. The famous Noble
well, which in 1863 flowed twenty-five hundred
barrels daily, and still flows largely, is on the same

side of the Creek as the Gillette Company wells, a little above. There also is the Crocker well, struck in 1863, and flowing largely for a considerable time, but now pumping.

Here became visible the usual system of transportation adopted for oil and fuel, which is flatboating on the creek. Four horses abreast are attached to a flat-boat, which they haul up stream, the horses taking the middle of the creek. The bed of the stream is even and covered with loose flat shale rock, the water being up to a horse's belly. An Oil Creek flat-boat generally holds from eighty to one hundred barrels of oil, on which the freight up is from seventy cents to one dollar, freight on coal being in proportion. As the boats sometimes make two trips a day, the business is highly profitable, though anything but pleasant, especially to the horses. As we passed down the creek the weather was intensely cold, and the ice was floating down in large masses, but the unhappy horses had to wade up with their heavy loads, their bodies partially clad in icy coats of mail, and their tails mere bunches of icicles. If it is borne in mind that these horses had to be from three to four hours in this icy water, without relief or rest, and that even saddle-horses have to wade the stream several times in making the journey, the short lives and the wretched character of the live-stock in that region will not be wondered at.

Passing one or two "runs" with derricks going up or wells going down, we strike the McElhinney Farm, on both sides of the creek, punched as full of holes as a strainer. Here is the famous Funk well, the first flowing well on the creek, that kept up its stream of wealth for fifteen months, and, close by, the Empire well, that gushed forth three thousand barrels daily, and flooded the land around with oil a foot deep.

The Funk well is now silent and its lips dry, but the old Empire, after two years of steady flow, followed by a pumping yield of about a hundred barrels daily, and then a stubborn refusal to give another drop, has been induced, by the gentle persuasion of a "blower," or air pump, to send up about a hundred barrels a day.

Passing through the Boyd Farm, on the East side of the Creek, on which there were fewer wells than on the tracts on either side of it, we crossed the stream to the G. W. McClintock Farm, where the throng of derricks, the clustered houses, and the flag-pole in front of a tavern, marked the presence of "Petroleum Centre." Here the wells are crowded as thickly as houses in the most populous part of a city, dwellings and engine houses being mixed up in such inextricable confusion that it is difficult to distinguish one from the other without entering, and not always then. A ravine enters the Creek at this point from the West, and near the mouth are several producing wells, among others the Wild Cat wells, on the Gillespie property.

Re-crossing the river to the East side we came out on the Hyde and Egbert Farm, one of the most noted parts of the Creek from the number of important wells on it, among them being the Coquette, which is now flowing a large fortune into the pockets of its owners. Like most of the original owners of property on Oil Creek, both Dr. Egbert and his partner were men of small means before they struck oil, though both are now very wealthy. Several of the great flowing wells of the Creek are on their farm, and as the land owners have an interest — generally one-half — in the oil raised, a comfortable income is secured. Here is the Maple Shade, which for months ran a thousand barrels daily, now dropped to fifty; the Jersey, which still flows three hundred and twenty; the four Keystone wells, which have aggregated two hundred and sixty barrels; and a number of other flowing wells, including the Coquette. There are, in all, on these forty acres, sixteen flowing wells, and sixteen pumping wells, all yielding.

The Coquette being one of the latest great "sensations," is an object of much curiosity, and

many pilgrims come daily to gaze in wonder and
envy on it. Steps have been erected and visitors
are admitted to a sight of the oil pouring into
the tank on payment of ten cents. A romantic
story is told in connection with this well. It is
said the brother of the superintendent of the
property had a dream in which he fancied that
an Indian menaced him with bow and arrow. At
that moment a lady friend, who had been considered somewhat of a coquette, advanced stealth-
ily and handed him a rifle with which he fired at
his foe. The Indian disappeared, and from the
spot where he had stood gushed out a river of
oil. Visiting his brother soon after, he recognized
a place on the Egbert Farm as the scene of his
dream, and pointed out the spot from whence the
stream of oil burst forth. His brother marked
the spot and bored the Coquette well, which
commenced flowing fifteen hundred barrels daily,
then fell to one thousand barrels, and now gives
about six hundred barrels, regular daily yield.
Owing to the volume of gas in the well the oil
rushes out like spray with such violence that at
first it blew entirely across the tank, and saturated the ground around. A covering of boards
has now been placed at the mouth of the pipe,
and against this the stream plays with a force
resembling the stream of a steam fire-engine
striking the side of a house. The quality of the

7

oil was at first not of the best kind, being "riley" and of a yellowish color instead of the dark green of pure oil. It has now greatly improved and sells at full prices. There are twelve tanks holding from twelve hundred to sixteen hundred barrels each, full of oil, worth in the aggregate over $150,000, in addition to what has been barrelled and sold. All around are notices warning visitors against smoking, the air being full of highly inflammable gas. A share in the Coquette is considered a "moderate" fortune. In January of this year Dr. Egbert sold one-twelfth interest in the Coquette for $250,000. Four years ago he bought the entire forty acres on which these flowing wells are located for one thousand dollars, taking his last dollar to pay the sum.

Passing the Rhinoceros well, the Porcupine well, the Ram Cat well, and a whole menagerie of other wells, we came to the Story Farm, crowded with derricks and wells. Here are the locations of the Columbia and Dalzell Oil Companies, two noted Pittsburgh Companies, the former being one of the most successful in the oil regions, returning larger profits to its original stockholders than any other company. Its history is such a remarkable instance of profitable investment that it will be read with interest. The Columbia was organized in 1862, and purchased the Story Farm for $128,000 from a company of seven persons, of Pittsburgh, who in 1859 bought the farm for a few thousand dollars. The Columbia Company was organized with a capital of $200,000, divided into 10,000 shares of a par value of $20 each. During the year 1862 the stock varied in price from $2 to $10 per share. At this time the chief difficulty with the company was the receipt of twelve hundred barrels of oil per day and no market for it. But a foreign demand soon sprang up, and between 1862 and 1864 the Company divided $300,000. In April, 1864, $70,000 was divided, in May following $100,000, and $100,000 in June. The dividends between July and December were $625,000, making a total of dividends since the formation of the company of $1,195,000, more than five times the amount of original capital. In June, 1864, the old shares were called in and new ones issued of $50 each, the holder of an original $20 share receiving five new ones, of $50 each. The person who paid one year and a half ago the par value of $20 each for one hundred shares, and has held his stock, has received $12,000 in dividends to December, and from the profits on the increase of capital made in June last, obtained an accession to his stock of four hundred shares, which shares are worth, with his original hundred shares, at present market prices, $42,500, making a clear profit of $52,500 in eighteen months. If he bought the original shares at their lowest price, $2 each, that profit was made on a capital of $200.

Next below the Story Farm, on the East side, is the Tarr Farm, on which is the famous Philips well, which flowed two thousand barrels daily for many months. The owner of the Tarr Farm in years past was a poor and uneducated man, who eked out a meagre livelihood by lumbering in addition to scratching the barren hill for a scanty crop. Poor as the surface crops may have been, the soil below has sent up products so rich that the lucky owner is now an exceedingly wealthy man, who lives in splendid retirement at a small town not far from Meadville. The farm is covered with wells, nearly all of which, if not the whole, are successful. Judging from a superficial examination, this appears to be one of the most successful territories on the Creek, as it certainly is one of the most muddy. Here we came across a team, stuck in a mudhole, the fore-wheels clear under and the hind-wheels invisible to the hub. The teamster, who, judging from that portion of him above ground, was probably a six-footer, stood contemplating the situation with dismay. In passing we ventured to remark, "Mister, guess you are stuck." It was a daring remark to make under the circumstances, and nothing could be expected in response less than a volley of curses, deep and dire. That such a proceeding suggested

itself to the mind of the teamster was evident by
the look of his eye, but, after revolving the whole
matter, he concluded he could not do justice to
the subject, and with one look at the "stalled"
team and another at us, he gave a heavy groan,
and responded, "Well it looks like it!"

Through muddy flats and up steep hill-sides;
past throngs of derricks; by gushing flowing
wells, and creaking pumping wells; through the
Blood Farm, where the dilapidated, unpainted,
moss-covered and time-stained house, in which
the owner of the farm lived in his days of poverty,
is confronted by the smart and showy boarding
house erected for the use of its employees by the
New York company now owning most of the
wells, at last we reach the Rynd Farm, and the
mouth of Cherry Tree Run. Here the wells
again become very thick, and abundant evidences exist of a large number of them being
productive.

On this farm are two peculiar wells, the "Agitator" and the "Sunday." The Agitator has to
be pumped for a few strokes every half hour,
when it flows for ten or fifteen minutes. The
"Sunday" well, close by, produces nothing on
week days, when the "Agitator" is worked, but
on the Sabbath, when the "Agitator's" pump
rests from its labors, the "Sunday" flows about a
couple of barrels.

7*

The widow McClintock, or Steele Farm, lies below the Rynd Farm. A large number of valuable wells are on this property, yielding a splendid revenue to the proprietor of the land, John W. Steele, who has but just come of age. The Buchanan Farm comes next in line, and at this point, near the village of Rouseville, another of the villages born of the oil excitement, is the old Taylor well that once flowed largely, but stopped after yielding sixteen thousand barrels, when it was left to lie idle. It has recently been restored by pumping to about thirty barrels. A few rods farther down, Cherry Run enters Oil Creek. Resisting for the present the desire to explore this ravine, now the scene of so much excitement on account of the numerous successful wells recently struck, we passed on to McClintock-ville, on the Hamilton McClintock Farm, a small village, partly perched on a high bluff, and partly on the low ground on the other side of the Creek.

Here again was a throng of wells, most of them highly successful, with several new wells, many of which had "struck ile." Among these wells was one which has been the scene of a curious streak of luck. The owner sank his hole to the third sand-rock, but found nothing but water, He pumped diligently for days, but without finding a grease spot in his vats, and then abandoned his unproductive hole in disgust. At this point he was visited by the owner of a neighboring well, who had been reaping the benefit of his labors, the water drawn up from the unproductive well having relieved the adjoining well of the stream which had previously caused some trouble. The visitor offered to pay the unfortunate pumper thirty-eight dollars per week to keep his pumps going, and, rather than abandon his engine, the latter agreed, and set to work once more. Six months he kept at it, drawing pure water out of the hole, to the relief of his neighbor, and then he "struck ile," and has since been pumping steadily to his own delight and the chagrin of his neighbor, whose vein he has "tapped." A nice question of ethics is involved in this matter. If the pumper was hired to pump his neighbor's water, has he any right to pump his oil?

In the middle of the river below the McClintockville bridge, is an old well. Tradition says that at this point a spring of oil bubbled up, and the Indians were in the habit of coming there to skim the oil for medicinal purposes. Here, also, it is said the owner of the land gathered the oil by soaking a blanket in the stream and wringing out the oleaginous fluid in a bucket. A few years ago a well was sunk on the spot, but the brilliant expectations indulged in by the adventurers were never realized. It was not a paying investment.

The Clapp Farm has a number of wells, many of them successful, but none of great note. Just below the southern line of this tract is Cornplanter Run, coming in from the West. Preparations have been made for boring on that territory. The Graff Hasson Farm, next above Oil City, contained one thousand acres, and was purchased in 1856 for $7000. A short time since three hundred and twenty-five acres of it sold for $750,000. It formerly belonged to Cornplanter, the renowned chief of the Seneca Indians.

Oil City at last. Oil City, with its one long, crooked and bottomless street. Oil City, with its dirty houses, greasy plank sidewalks, and fathomless mud. Oil City, where horsemen ford the street in from four to five feet of liquid filth, and where the inhabitants wear knee-boots as part of in-door equipment. Oil City, which will give the dirtiest place in the world three feet advantage and then beat it in depth of mud. Oil City, where weary travellers think themselves blest if they can secure their claim to six feet of floor for the night, and where the most favored individual accepts with grateful joy the offer of half a bed and the twentieth interest in a bed-room.

Oil City is worthy of its name. The air reeks with oil. The mud is oily. The rocks hugged by the narrow street, perspire oil. The water shines with the rainbow hues of oil. Oil-boats, loaded with oil, throng the oily stream, and oily men with oily hands fasten oily ropes around oily snubbing-posts. Oily derricks stand among the houses, and the "town-pump," if there is such an institution, must pump oil. There are several productive wells in the city, ranging from five to twenty barrels, and the citizens are busy boring in their back yards, in waste lots, or wherever a derrick can be erected. The Linden well, just above the Petroleum House, is remarkable from the fact that it commenced to flow on the 10th day of October, 1861, at the rate of twenty barrels per day, and has daily yielded a supply that has not varied five barrels during the whole period, and appears to be as vigorous to-day as when first struck.

The growth of Oil City is something remarkable. Until the commencement of oil mining on the Creek, there was nothing at the junction of Oil Creek and the Alleghany but a small store and a tavern or two, frequented by the raftsmen who brought their rafts into the eddy and rested awhile. In 1861 a settlement was established at the mouth of the Creek, and several stores of various kinds put up. In the Spring of 1862, Oil City was incorporated as a borough. There are now nearly one hundred stores and work-shops of various kinds, to which additions are always making. Fourteen hotels, large and small, and half a dozen saloons, minister to the bodily comforts, whilst the spiritual wants are supplied by four churches. Healthy as Oil City is claimed to be, there has been found employment for eight doctors, who, however, frequently mingle oil speculations with their practice.

In travelling from Shaeffer's Farm to Oil City, and not taking into account any of the " Runs," over one thousand wells, old and new, are passed. In a short time that number will be largely increased.

CHERRY RUN.

THE fact that Oil Creek itself is not the only valuable oil producing locality, and that apparently valueless territory may prove highly productive, has been exemplified in the history of Cherry Run. A year ago this property was almost entirely neglected, very few derricks were erected on it, and the land was held at comparatively low prices. Now there seems to be no limit to the sums asked and paid. The principal cause of this excitement is the success of the Reed and Criswell and other flowing wells which "struck oil" on the Run during the Summer and Fall of 1864. The Reed and Criswell well commenced flowing about a thousand barrels daily, but soon dropped to two hundred and eighty barrels, at which it has remained steady for several months. The quality of the oil is very fine. Soon after the striking of the Reed well, some others commenced flowing. The excitement became intense, the rush was tremendous,

and in a short time all the available property on
the Run was taken up at high figures. Failing
to secure the fee-simple to the land, the next
object of the late comers was to secure leases,
and in order to obtain these the anxious oil-
seekers were not only ready to give half the oil,
but to pay large bonuses in addition. Soon the
valley was planted as thickly with derricks as
it could possibly hold, where a lease could be
obtained, and even the steep hillsides were bored
by the pertinacious oil-seekers. To the utter
confusion of theorists who hold that oil can only
be found on the flats, and to the triumph of those
who hold the opposite opinion, several of the
wells away up the hillside have proved successful, which furnishes another proof that the only
reliable theory is that oil exists wherever it flows
or can be pumped out of a well, in other words,
where it is it can be found, and where it is not it
will not pay for seeking it. Nothing additional
is charged for this bit of valuable information.

The scene in going up Cherry Run is more full
of excitement than anywhere on the Creek. The
Run, at its mouth, is but a narrow gorge between
steep hills, through which a muddy, rocky, and
brawling stream wends its way. Pumping wells
and flowing wells are planted thick along the
narrow flat, and climb the hills on either side.
The road is execrable at its commencement, the

wagons sinking over the hub, and, at a short distance from the Creek, loses its identity in a number
of deep ruts looping around in all directions where
a teamster could force his team in hopes of finding
a better track.   The stream being small and the
bed rocky, water transportation is not available,
and powerful teams, drawing wagons loaded with
five barrels of oil each, go plunging and stagger-
ing in the mud, and among the rocks that form the
bed of the stream.   Every now and then a wagon
breaks down, and then the perpetual chorus of
shouts and oaths becomes intensified in spots,
making, with the noise of escaping steam, the
clank and jar of the engines and pumps, and
the rushing of the stream, a noisy entertainment.
Travelling on the Creek is bad enough, but the
extreme of diabolical locomotion is not attained
unless the tramp is taken up Cherry Run.   The
constant passage of teams not only cuts the roads
into deep sloughs of mud, but makes the pedestrian keep bobbing around to escape being
knocked over by them in their erratic courses.

About two miles up the Run is the Reed and
Criswell, or Reed well as it is generally known,
Criswell having sold out his interest for a princely
sum.   In the vicinity of this well are several
other flowing wells, among them the Baker well,
credited with one hundred barrels daily ; the
Gruninger, an intermittent flowing well ; the

8

Yankee, flowing fifty barrels; and, a short distance above, the Auburn, flowing eighty barrels. The two acres, on which the Reed well is located, were offered for sale two years ago, for $1500, but found no purchasers. It was lately sold for $650,000. Next above the Reed well is the Smith Farm, comprising fifty acres. Three or four years ago the then owner offered it for sale at $250 over the incumbrances. It was afterwards sold for $2,400, and resold a year since for $6,500. It is now the property of the Cherry Run Oil Company, who have done nothing of themselves to develop their property, and have therefore been at no expense beyond the original purchase, but who are receiving from leases on the territory over four hundred barrels of oil daily, in royalty. A new well was struck a short distance above the Reed Well in January, flowing two hundred and fifty barrels daily, without the sucker rods being pulled out. Beyond the Smith Farm is the McFate Farm, on which there are several wells either down or going down.

Here the region of producing wells may be said to terminate for the present. Above this point derricks innumerable are planted along the valley and hillsides, but engines are scarce. Leases are taken and derricks erected thickly to nearly ten miles from the mouth, but nothing further has been done. The terms of most of the leases made on this Run require that derricks shall be erected within sixty days, and engines be on the ground within ninety days from the date of the lease, and that the work be then prosecuted with all reasonable diligence. In order to secure the lease as far as possible, the derrick is in all cases erected within the required time, but considerable difficulty exists in getting the engines on the ground, owing to the great demand on the machinists for engines.

Nearly four miles up the Run is the Humboldt Refinery, a very extensive establishment. The shipping point of this refinery is on the Alleghany in Walnut Bend, and in order to facilitate transportation the proprietors have constructed a road over the mountains, at a considerable expense, and established a ferry across the river. Most of their crude oil is brought from the creek in wagons, but a considerable quantity is pumped up in pipes from the Tarr Farm.

Plumer village lies a short distance beyond the refineries, and about four miles from the mouth of the Run, or seven miles from Oil City by the road. The struggle to get property on Cherry Run has been so eager, that land has been purchased or leased for boring purposes a considerable distance above. At the beginning of the year 1865, there were on Cherry Run, from Rouseville to Plumer village, one hundred and eighty-six derricks, and about fifty above Plumer.

CHERRY TREE RUN, WEIKEL RUN, CORNPLANTER RUN, REED RUN AND TWO MILE RUN.

THE Cherry Tree Run enters Oil Creek from the Northwest, on the Rynd Farm. It is a ravine of considerable size, with abrupt and lofty banks near Oil Creek and widening into a fine valley as it approaches the quiet little village of Cherry Tree. Until very recently no attention was paid to this valley as an oil locality, but the great success of the investigations on Cherry Run led to a more careful examination of all the ravines in the neighborhood of Oil Creek. There are now a number of experimental wells going down, nearly as high up the Run as Cherry Tree Mill. None of the wells have as yet reached a sufficient depth to fully test their value, except one struck in March of the present year, which is reported as yielding one hundred barrels daily. As with all other partially developed territory, the companies engaged in boring are very sanguine that Cherry Tree Run will prove

(88)

as much of a success as anything in the oil regions.

Weikel Run branches from Cherry Tree Run a short distance above its mouth. It is a narrow ravine with steep banks, covered with timber. It has lately become favorite territory from the belief very largely entertained by experienced managers of oil wells that large deposits of oil exist in it. The indications certainly favor this idea, the configuration of the ravine giving it the appearance of a slightly diminished copy of Cherry Run. Less than half a mile up is the celebrated "great gas well," which made such a noise in 1864 by its unprecedented explosions of gas. The gas vein was struck in May, when it blew with such violence as effectually to put a stop to all further attempts at working. The volume of gas was tremendous, and its violence so great that anything thrown on the hole was instantly jerked into the air. Its roaring could be distinctly heard for a hundred rods. For nearly six months it continued to blow gas without cessation or apparent diminution, until at last the hole was plugged after considerable trouble. As soon as it was plugged the gas forced a stream of water into a well sunk some distance up the Run.

In December the gaseous manifestations ceased, and the proprietors are preparing to sink the well

8*

with the expectation of finding a large deposit of oil. Whether their expectations will be realized remains to be seen, but on the principle that there can be no smoke without a fire there is evidently considerable oil in the neighborhood, if not immediately under the well. Acting on this belief the land around has been taken up by companies who propose testing it thoroughly. A small patch of land near the well, containing a small strip of borable territory and the rest "set up edgeways," has been purchased for $9000 by a Philadelphia company, and immediately above the gas well is the property of the "Weikel Run and McElhinney Oil Company."

A short distance beyond the gas well is another well going down, having reached 660 feet. The show of oil was very encouraging, and the borers were confident of finding oil in considerable quantity, as soon as the third sand-rock was reached. Other wells in various stages of progress are scattered along both sides of the stream, considerable activity being displayed in developing the property.

Taking the road from Cherry Tree to the Alleghany River, Weikel Run was left behind, and we crossed Cornplanter Run, which enters Oil Creek on the Clapp Farm. Preparations are making along its banks and bed to bore in the spring, but as yet no developments of consequence have been made. A little beyond this point we left the road and came out on Reed Run, a branch of Two Mile Run, which comes into the Alleghany not far from Franklin. Here were abundant evidences of a determined search for oil being in progress.

Reed Run, is a very attractive territory to those who have a good eye for desirable oil locations. The bottom land of the Run, and the second bottom a few yards higher, afford as good prospect for successful wells as can be found on any of the Creeks or Runs in the vicinity of Oil Creek.

Near the Junction of the Reed with Two Mile Run, is a well, down 350 feet, with good indications of oil.

Two Mile Run has not yet any producing wells, but preparations have been made for boring it thoroughly.

THE ALLEGHANY RIVER, WITH HICK ORY, TIONESTA, HEMLOCK, PITHOLE, AND OTHER TRIBUTARY CREEKS.

DURING a portion of the summer months, before the oil excitement extended far up the river, a small steamer occasionally ran up from Franklin to President, on the Alleghany River, but those desirous of going higher up, had to seek some other mode of conveyance. In the Fall and Winter months, no boats run above Oil City. A road follows the course of the river, with ferries at the points where the jutting of the precipitous bluffs out into the stream stops the way. To those unaccustomed to the region of rapidly flowing rivers, these ferries are interesting novelties. Two strong and lofty poles are firmly fixed in the banks and across their tops is stretched a stout wire or iron rod, the ends fastened to the rocks behind. A "traveller" or pulley wheel is placed on that part of the wire which crosses the stream, and from this "travel-

(92)

ler" a line passes to the ferry-boat, which is a flat, clumsy affair, on which passengers, horses and vehicles are jumbled together indiscriminately. When a load is on board, the boat is pushed out into the stream, and the force of the current carries her over without paddling, or care of any kind, the "traveller," in its passage across the wire, emitting an eerie sound that echoes strangely among the wild hills at evening, and proclaims to those who have ears to hear, the need of a well of lubricating oil in the neighborhood.

Those wishing to strike the river above Walnut Bend, can shorten the distance considerably by taking the road from Oil City to Plumer, about seven miles, and thence by way of Neillsburg to Tidioute, or striking across to President, reach the same point along the river road. The latter route is the most interesting, and generally is a better road for travelling.

A daily stage runs from Oil City to Plumer, but beyond that there is no public conveyance. Unless capable of performing a long march on foot, the best course is to get a horse in Oil City, and set out early in the morning. It is between forty and fifty miles to Tidioute, and there being much to see, and Oil City horses not noted for speed or bottom, the hours wear away rapidly.

Riding up the steep hill-side from Oil City, a

fine view is obtained of the lower part of Oil
Creek with the crowd of wells on the broad flat
through which the stream runs towards the
close of its course. A good bird's-eye view could
be taken from this point, although not represent-
ing the busiest portion of the oil region, and
some enterprising photographist may find this
hint profitable.

About a mile before reaching Plumer, the road
crosses Cherry Run, and the multitude of derricks in the valley and along the hillsides, testify
to the favor in which the Run is held by oil adventurers. The Humboldt Refinery lies in the Run,
to the left of the road, and all the details of its
extensive area lie open for inspection as on a
builder's plan. The road at this point is horribly
cut up by the heavily loaded wagons conveying
oil to and from the refinery, and is very nearly
as bad as the principal street of Oil City. Climbing another low hill the village of Plumer is
reached, and at the centre of the village the road
to President branches off to the right.

Very little can be said in favor of the road
from Plumer to the river. The first part is bad
enough, but on reaching Pithole Creek it becomes
worse. The creek well deserves its name, as it
winds its way through a gorge, dark, deep and
forbidding. The road winds along the face of
the precipitous sides, the last part of the descent

being very steep and exceedingly miry. Long before reaching the bottom the rushing sound of the waters can be heard. A mill spans the stream, the road being carried by a narrow bridge across the mill dam, and climbing the other side amid crags and boulders at so steep a grade that a firm seat and steady hand are necessary in making the ascent or descent.

Pithole Creek obtains its name from some holes or small caverns in its sides from which a mephitic gas arises. A dog held close to one hole expired in a few minutes, and a goose, put into the hole, died in three minutes and soon became corrupt. A stone thrown into one of these holes can be heard rattling from ledge to ledge in its descent, until the sound dies out, rather than stops. The existence of these gas exhaling caverns led several persons to sink wells in hope of finding oil, but, although some success was met with at the mouth of the creek, no very encouraging results were obtained higher up until, about the middle of January, 1865, a well on the Holmden tract, about seven miles from the mouth of the creek, and about five miles northeast of Plumer, in an undeveloped territory, struck oil and flowed at the rate of two hundred and fifty barrels per day. This created an immediate excitement, and Pithole Creek was swarmed with speculators eager to buy or lease every rod of land in the vicinity. The success of this well has demonstrated the fact that large supplies of oil can be obtained above the line of Oil Creek, and has increased the expectations of up-river oil seekers. The surface level of the Frazier well, is seven hundred feet above the level of Oil Creek, and the oil was found one hundred feet above the surface level of Oil Creek.

A peculiar circumstance connected with this Pithole well is the fact that it struck oil in what is known as the fourth sand-rock, being the only well in the oil regions that has reached that stratum. The first sand-rock was reached at one hundred and fifteen feet, the second at three hundred and forty feet, the third at four hundred and eighty feet, and the fourth at six hundred feet. At six hundred and eight feet oil was struck. The well was drilled in November, but was not tubed until the middle of the following January. As soon as tubed the pump was set to work, and after an hour's pumping the oil began to flow, with the sucker-rods in the chamber, at the rate of two hundred and fifty barrels per day, at which rate it has steadily continued. The well is the property of the United States Petroleum Company.

After leaving Pithole Creek there is a good road on either side of the river, all the way to Tidioute, the hills falling back, or being less abrupt as the ascent of the stream is made, and better farming-land appearing on the bottoms and in the rifts of the hills.

Tidioute, in Warren county, is the highest point on the Alleghany where there are producing wells. Above that point there are several of what are known as "farmers' wells," sunk in 1860 and 1861 by hand to a shallow depth, and abandoned when the depression in oil affairs occurred. Some attention has again been attracted to these wells, and preparations have been made for sinking them deeper, and also for testing the oil-producing qualities of Big Broken Straw Creek, which enters the Alleghany above Tidioute.

The Economite wells, owned by a religious sect known as the "Economites," are nearly opposite Tidioute. Five producing wells yield an aggregate of about sixty barrels daily, of heavy oil, the depth of none of the wells being over one hundred and twenty feet. The Economite wells are in the side of the steep bluff a little way up from the river. The "Brethren" are putting down several new wells in similar locations. On the flat across the river, some other parties have put down a well six hundred feet without obtaining oil. The Tidioute and Warren Company, in the immediate neighborhood of Tidioute, and also on the side hill, have recently obtained a pumping well of a hundred and fifty barrels per day, at a depth of one hundred and forty-six feet, of which eleven are in the first sand-rock.

About half a mile below is a well down a nearly a thousand feet, belonging to the Tidioute and Alleghany Company. Only one sand-rock has been passed through, at a depth of one hundred and fifty feet. A good show of oil and gas has been obtained, but the deposit of oil has not yet been reached. On the other side of the river are some shallow wells, of the same depth as the Economite wells, producing five to six barrels daily of lubricating oil.

Entering Venango county, the first point where active operations have been commenced is at the mouth of West Hickory Creek, on the upper part of what is known as the Hickory Town Flats. Several companies have located on these flats, and are sinking wells, with varying prospects of success.

Close to the mouth of the creek, three Scotchmen, named McKinley, sank a well in 1861. They reached a depth of two hundred and thirty-three feet and found a fine supply of oil, promising to yield them a rich return for their investment and labor. Just as they were making preparations for tubing it the war commenced, and the owners of the well became so much alarmed at the condition of affairs that they abandoned their undertaking, and their lease became forfeit.

On the other side of the river, just opposite the McKinley well, is a well down 340 feet, tubed, and ready to pump, with a fine show of oil. The intensely cold weather has prevented further operations. On East Hickory four wells are going down. About a hundred rods below the lower well of the Hickory Farm Company a Pittsburgh company are putting down a well. All of the adjoining lands have been taken up at high prices, and in the spring there will be a large number of wells sunk in the neighborhood of the Hickory.

About a mile below West Hickory Creek, on the same flat, is the property of the Pittsburgh and Alleghany Valley Oil Company, covering about two hundred and twenty-nine acres. In 1861 a well was sunk on the property to the depth of two hundred and twenty feet, with a good show of oil, but was abandoned on account of the low price of the product. As yet no work has been done on the property by its present proprietors beyond the necessary preliminaries for commencing operations, but the engines will soon be on the ground, and the work of properly developing the tract commenced.

About half a mile below the Pittsburgh and Alleghany Company's tract is the Sowers Farm, on which is a well that was struck in 1861 and flowed largely, but which, like nearly all the wells in the country, was abandoned in consequence of the low price of oil. It has now been put in operation and rendered productive.

Tionesta Creek comes in from the East a short distance below the Sowers Farm, which lies on the West bank. A few scattered houses and a tavern fronting the ferry landing, form the village of Tionesta. Around the mouth of the Creek and along the banks of the river there are abundant evidences of oil speculation, past and present. Shallow wells, hastily abandoned in 1861, rear their time-stained derricks on every side, whilst workmen, busy getting out timber for new derricks, and eager, keen-eyed men, with travelling pouches strapped to their sides, out "prospecting" for desirable sites, show the revival of interest in oil matters. A number of islands stud the surface of the river from Tionesta, past Lower Tionesta Creek, down to within about a mile of President, and on many of these islands old derricks and new derricks rear their heads among the unshapely trees.

The village of President, with its large, new, smart hotel, and its respectable gathering of houses, marks the junction of Hemlock Creek and Porcupine Run with the Alleghany; above the village is the well of the Farrar Oil Company.

The whole territory surrounding the village, and extending up Hemlock Creek and Porcupine Run in one direction, and down the river a considerable distance in the other, covering in all 8,400 acres, is the property of the President Petroleum Company, probably the largest corporation yet in the oil field, having a capital of five million dollars. Three wells are on the property, one near the McCrea Run, having reached a depth of 400 feet, with good show of oil, and two others on the river front, just below, going down on a lease taken by the Heydrick brothers.

Like Pithole Creek, Hemlock Creek has some strong manifestations of gas, or mephitic vapor. A story is told of three young men going along the valley in winter and finding the snow melted around a hole in the ground. One of them, a notoriously profane fellow, swore it was an opening into hell, and that he intended warming his feet at the fire. His companions endeavored to dissuade him, but he sat on the ground and stuck his feet in the hole, swearing with horrible oaths that he would warm his feet there if he had to go straight to hell in order to do it, and thanking the devil for finding him such an opportune supply of fuel. In a few minutes he stopped talking, and when his companions dragged him away he was totally insensible from the effects of the gas. His recovery was very difficult.

9*

Just below the President Company's tract, at the foot of a lofty bluff, is the celebrated Heydrick Well, sunk three years ago by the Heydrick brothers, young men who lived on the land, and who early adventured in grease. The well flowed for a considerable time from four to five hundred barrels daily, and then pumped one hundred barrels. When oil fell to a mere nominal value, and an empty barrel was worth its contents in oil six times over, the well was allowed to remain idle. It has now been started up again and is making from twenty-five to thirty barrels daily. The farm was owned by two Heydrick brothers and a brother-in-law, who, like most of the farmers in the oil regions, had enough to do to make both ends meet. The well sunk on their lands has been leased by the Farmers and Mechanics' Company, who pay the Heydricks half the oil, a tribute which puts a comfortable sum daily in their pockets. Four years ago a big flowing well was gushing out oil next to the Heydrick property, under the management of a Michigan company. When the Heydrick well was struck the Michigan well stopped, and no attempts have, so far as we could learn, been made to recover the vein. The Heydrick well has flowed and pumped, to the present time, over thirty thousand barrels.

On the west bank of the river, directly opposite the Heydrick well, is the Henry Farm, the pro-

perty of Hussey and McBride, on which there are
several productive wells. One has been flowing
and pumping with large returns for two years,
and now yields forty barrels a day. Another
" struck oil " at the depth of 400 feet, and is
yielding a hundred barrels a day. Still another
recently struck oil and is giving large returns.
But the principal well was sunk in 1861, and, at
the depth of 242 feet, obtained a flow of oil that
bewildered the proprietors. The greasy fluid
gushed up at the rate of three hundred barrels a
day, and continued to flow for three months.
What to do with the oil was a puzzle. Barrels
could not be got to ship it to market, nor vats to
hold it on the ground. Oil was down in the
market, bringing but ten to fifteen cents a barrel.
There were no refineries in the neighborhood, and
like the man who won the elephant in the raffle,
the proprietors of the flowing well were "put to
their trumps" to know what to do with their
prize. A little ravine close by was dammed up
and the oil turned into it until about an acre of
pure oil covered the ground two or three feet
deep. At last, in despair, the tube was stopped
with a pine plug, but the grease oozed up and
escaped. The tube was then bound over with the
never failing seed-bag, but the oil burst through
the earth and escaped into the river. Since then
the well has remained plugged, but it is now to

be deepened and re-tubed, when it is expected the old vein will be again reached.

Next above the Henry Farm, and lying across the mouth of Culbertson Run, is the property of the Beekman Oil Company, on which there are three wells, having a good yield of oil. Beyond this, on the McCrea Flats, is the Kelley well, now owned by the Cleveland and Buffalo Petroleum Company. This well pumps from twenty to twenty-five barrels of oil daily.

Near the property of the Cleveland and Buffalo Company, on Culbertson Run, is that of the McCrea Petroleum Company of Pittsburgh. The tract leased by the McCrea Company is partly on the opposite side of the Run to that of the Cleveland and Buffalo Company, and also overlaps the stream for a short distance. A well was sunk on the property in 1861, reaching 280 feet with a fine show of oil, when the break-down in the price of oil occurred, and the well was abandoned. Two engines have been purchased and are on the ground, to sink new wells, besides drilling the old one deeper.

On the McCrea Farm, lying between the property of the Beekman Company and the lands of the McCrea Company and Cleveland and Buffalo Company, is the territory of the Eagle Company, of Philadelphia, which "struck oil" during the last week of January, in one of their wells, which is now running over fifty barrels daily. This strike has made the property of other companies in the neighborhood increase greatly in value.

Past Pithole Creek, with the numerous wells clustering around its mouth, some of them producing a fair yield, down to Walnut Bend and Walnut Island. All along, the numerous spires of smoke from engine houses, the creak and wheeze of engines, and the steady plash in the black and greasy vats, told the story of remuneration for faith and labor. On Walnut Island a hundred-barrel well was struck early in January this year, and gives signs of increasing.

From this point down the bluffs increase in height and steepness, and the flats are generally of less extent. Derricks line the narrow path at the foot of the bluffs, sometimes climb part of the way up the sides, and are planted thickly wherever there is a moderately wide shelf, or where a stream makes an opening in the hills. At Horse Creek, on the East side of the river, the Ross Oil Company's well is pumping about twenty barrels per day, and some other wells on that side are doing more or less, among them being the Wheeler well, doing thirty barrels. On the West side of the river, after getting a short distance below Walnut Bend, is the Hulings well, pumping twenty barrels. The Phillips well, unproductive, having stopped its yield some time since; the managers are putting down a new well. The Revenue well, opened three years ago, abandoned, and now running, under new management, about twenty-five barrels a day. The gas from this well is used to save a part of the fuel in the engine fires. About opposite Horse Creek is the Brady Bend well, formerly flowing, now pumping; College and Kincaid and Porter wells, four years old, and now revived and pumping each eight barrels per day. Kincaid new well, that pumped sixty barrels per day for the first three days of its working in November, and now doing finely. Harrington well, pumping twenty-five barrels.

Farther down stream is the well of Long & Gay, pumping at a depth of 530 feet, with a fair yield. Beyond are the wells of Purchase & Co., two wells sunk in 1861 and abandoned, now cleaned out and pumped with one engine. One well commenced pumping in September, 1864, at the depth of 348 feet, yielding fifteen barrels, with considerable gas. The other was pumped two months later, and gives twenty-two barrels, from a depth of 517 feet. Five minutes' walk distant is a well down 530 feet with a fair show of oil. The derrick bears the legend of "Oil, or China," and the borer swore he would either raise the oil or send his drill "up" through some Chinaman's cellar floor. Still nearer Oil City is the Alcorn Farm, on which are the two wells leased by the Cleveland Cherry Valley Company. One of these wells is down 600 feet, and has commenced pumping oil, with every prospect of yielding in the neighborhood of one hundred barrels daily as soon as properly worked. The other well is down 200 feet, with a good show of oil. These wells will undoubtedly prove a valuable adjunct to the other property of the company, and enable them to pay good dividends on the stock.

From this point to Oil City, about one mile, there are a number of old wells and new wells, the latter just commenced to go down, and some of the former recently cleaned out and prepared to be sunk deeper.

From Oil City to Franklin is seven miles. The scenery along this part of the Alleghany river differs but little from that above Oil City, which has already been described. Where the bluffs approach the river they tower up to a considerable height, rising abruptly from the water, and having their craggy sides partly covered with timber. Where the bluffs recede, there is, between them and the river, a strip of tillable land, sometimes a quarter of a mile wide, and then narrowing to a mere ribbon, which is at length terminated by the steep bluffs. These intervals of low land generally consist of two levels, one but little above high water mark, and the other a plateau from ten to thirty feet above.

All along the river-bank, on both sides, are oil wells, some of them yielding successfully, and others not yet sunk to the oil basin. Most of the wells are sunk on the strip of low land immediately adjoining the river, but a few are on the plateau, and several along the base of the steep bluffs or on ledges a few feet up the face of the bluff. Several of the wells gave good evidence of a fair yield, the stream of oil being of paying size and good color. Few have as yet properly developed their property, not having gone down to the third sandstone, which here lies deep, but contenting themselves with the yield from the less productive second sandstone.

The fact that all the wells along this part of the river are only down to the second sandstone, makes it evident that they cannot give as large a yield as the great flowing wells on Oil Creek, that are down to the third rock. But the corresponding fact that the second rock wells all yield lubricating oil, commanding more than double the price of the Oil Creek product, is a complete offset to the smaller yield. By boring from thirteen hundred to two thousand feet, it is believed that the third sand-rock can be reached, with a greater flow of oil than can now be obtained on Oil Creek. The correctness or error

of this assumption will be tested before long, as,
lower down the river, some well-owners think of
sinking two thousand feet, if the third rock is
not found at a less depth. The fact of the superior facilities for shipment possessed by wells
on the river, especially within the limits of
regular steamboat navigation, is too self-evident
to need argument. It is enough to add that the
expense of getting the oil to the place of shipment is always taken out of the price of the oil
at the wells, and that oil produced on the river
will always therefore bring a higher price than
the same quality produced at a less accessible
place.

Passing several wells in operation, old wells
being cleaned out and prepared for re-working,
and new wells boring, Franklin was at length
reached, the quaint old capital of Venango
county, with its old-fashioned houses, its muddy
streets, and its miserable tumble-down Court-
house, in which land-sales of from one to three
million dollars a day are recorded, and documents
of incalculable value are stored without a vault
to protect them from the accident of fire. The
average consumption of revenue stamps in the
Recorder's office is estimated at about $500 per
day, making a very handsome revenue to the
government.

Franklin is a very old settlement, being the
10

site of three forts, Fort Venango, established by
the French, a fort built by the British, and Fort
Franklin, built by the Americans in the war of in-
dependence.   In 1795, the town of Franklin was
laid out on the site of the last named fort, and
afterwards became the capital of Venango county.
It now contains a population of about thirty-
five hundred, and is a growing place.   A suspen-
sion bridge spans the river at this point, the old
bridge having been burned down over eighteen
months since, by some blazing oil-boats that took
fire in the great conflagration at Oil City.   Franklin is the present terminus of the branch of the
Atlantic and Great Western Railway, over which a
very large business is done.   The Jamestown and
Franklin Railroad, partly built, will also connect
Franklin with Cleveland and Buffalo, by way of
Ashtabula and the Lake Shore Railroad.

There are several wells in operation within
the borough limits, and the product is very satisfactory to the owners.   Below Franklin the river
is lined with wells for several miles, many new
ones going down, and several old ones flowing
and pumping.   Among the producing wells are
the Keystone Well, pumping about four barrels;
the Lee Well, about five hundred feet deep, and
flowing about fifty barrels; the Dale & Morrow
Well, pumping about thirty barrels from a depth
of four hundred and fifty feet; the Hoover-Island

Well, on the first island below Franklin, pumping and flowing seventy-five barrels daily. A num. ber of wells with fair yield of oil are scattered along the river to a distance of several miles.

On East Sandy Creek, which enters the Alleghany a short distance below Franklin, from the East, a well yielding one hundred and fifty barrels, has been struck immediately above the junction with the river. A number of wells are going down for several miles up the stream.

On Big Sandy Creek, which enters the river below from the West, there are several wells going down, one of which is reported to have struck oil in large quantity early in April of the present year. On the Big and Little Scrubgrass, lying still farther south, there are several wells going down.

FRENCH CREEK, SUGAR CREEK, AND OTHER OIL LOCALITIES.

FROM Franklin to Meadville, twenty-eight miles, the Franklin Branch of the Atlantic and Great Western Railway runs along the bank of French Creek, an important and pretty stream, considerably larger than Oil Creek during the greater part of its length, and also deeper. For several miles up the Creek there are old wells and new wells, several of the latter producing oil; among the most noted being one on the Sutley Farm, a short distance above Franklin, and the well of the Tallman Company, near Utica Station. The oil produced on French Creek, being of a heavy lubricating quality, bears a higher value than that of Oil Creek. The land along nearly the whole length of the Creek has been purchased or leased for boring, and most of the abandoned wells of 1861 have been taken by new companies who have the capital and energy to properly test the property.

(112)

A favorite region at the present time is Sugar Creek, which takes its rise in Cherry Tree township, Venango county, on the same tract out of which Cherry Tree Run flows. It passes through the Borders of Plum and Oakland townships, in a Southwest direction, to Cooperstown in Jackson township, and then runs nearly due South through Sugar Creek township to French Creek, which it strikes about two miles above Franklin. It receives several branches, the largest of which is West Sugar Creek, which rises in Sugar Lake, just over the Crawford county border, and joins the main creek at Cooperstown. Through its whole course it passes through a fine farming country, the cultivated flats and hill-sides and good roads affording in this respect a decided contrast to some portions of the oil regions.

The Creek is not large enough to afford water facilities for shipping oil, but a good road keeps along the flat valley to the junction with French Creek, four miles, where there is a station on the Franklin Branch of the Atlantic and Great Western Railway. In passing up the Creek, evidences of explorations for oil, and of the oil itself, become speedily visible. About a quarter of a mile above the junction with French Creek, on the Homan Farm, is the well of the Sugar Creek Company, of Philadelphia, struck in February of the present year, and yielding, it is estimated, about seventy

10*   H

barrels of lubricating oil. The strike of this well caused a rush to Sugar Creek, and the value of lands has increased enormously. Several companies are busy putting down wells. On the McCalmont Farm, about two miles and a half above French Creek, is a well put down about three years since, and worked by water-power. In all, from three to four hundred barrels of heavy lubricating oil have been obtained from this well, such as now sells at twenty to twenty five dollars per barrel at the well. The hole was only sunk to the second sand-rock, reaching a depth of three hundred and twelve feet; none of the old wells on this creek have gone to a greater depth or penetrated beyond the second sand-rock, the supply of heavy oil having induced the owners to stop at that point. A company from Rochester leased the well, put up an engine, and in a very short time struck a vein of pure lubricating oil.

From this point up there are several wells and derricks, but few producing anything of consequence as yet, until Cooperstown is reached. Immediately above the village, on the Booth and Hillier Farm, a well similar to that of the McCalmont Farm was put down three hundred and twelve feet by water power, when it struck lubricating oil. The well has been purchased by two experienced parties from Oil Creek, who are confident of finding a good supply of oil in the third sand-rock, toward which they are boring, having reached a depth of six hundred feet. Two wells are going down on the Sweeney Farm next above, one having got down one hundred feet, and the other three hundred feet, the latter striking oil at eighty feet.

Adjoining Cooperstown on the south, and partly bounded by the borough line, is the farm of the Sugar Creek Company of Cleveland, containing one hundred acres, a large proportion of which is good boring territory. On an island included in the property, is a well put down three years ago by hand, to a depth of two hundred and ninety-seven feet, with a good show of oil. When the panic occurred the well was abandoned. It has now been leased and will soon be sunk deeper and tubed. Another well is going down on the property, and preparations made for a third well. A company from Cleveland and Sandusky have purchased the Alexander Farm, next below the Smith Farm, and are preparing to sink wells.

Cooperstown was a quiet little village, until about two years ago, when it was the halting place for the teams carrying oil from Oil Creek to Meadville and other points on the Atlantic and Great Western main line. Then the quiet and order was rudely broken in upon. Long lines of wagons stretched along the streets. The taverns were crowded with jovial and noisy teamsters, and the gutters were greasy with oil. When the Franklin Branch Railroad was built the teaming across the country was stopped, and Cooperstown reverted to its former quietness until the sudden excitement on Sugar Creek a few months since. Now all is changed, and the demon of oil has unrestricted sway in the village. Land offices and petroleum agencies are to be seen on all sides, and nothing is talked of or thought about but oil.

In the "Historical Collections of the State of Pennsylvania," published in 1843, is an interesting note in relation to the Indian relics said to have been found in Cooperstown and its vicinity. The writer says, " Skeletons were dug out of the bank near the mill-dam in that place. The whole valley of Sugar Creek once sustained a dense Indian population. Tradition says that the French, a century ago, worked silver mines on the spot where the village of Cooperstown stands. When the dam was being erected for the mill they made quite an excavation in front of the place now occupied by the store of Fetterman & Bradley. Some six feet below the surface a quantity of charcoal was found, together with a furnace and smelting vessel. Several specimens of ore were obtained also. The vein appears to be under the bed of the stream, as a deep excavation has been made there. This tradition exactly corresponds with an idea I have for many years entertained, viz.: that an abundance of lead, and, perhaps, of the precious metals, will yet be discovered in this county. The Indians undoubtedly procured their lead somewhere in this vicinity. Indian chiefs have been known to take silver ore from this section to Canada and trade it to British merchants. An aged chief of the name of Ross confidentially assured an old citizen of this county, that there were metals found and mines worked by the Senecas. He and Black Snake, a Seneca chief, concur in stating that there were mines between this place and Conewango; one is about seven miles from Cooperstown; the second mine was near Pithole, not far from Mr. Holeman's. It is called 'Cushing,' from the Seneca word 'cush,' meaning hog."

From Cooperstown to Utica is but a short drive, and not a long walk. There the cars can be taken for Meadville, and thus the grand circuit of the Venango county oil regions be completed. The weary traveller will be glad to exchange the discomforts and hardships of his tour for the warm welcome and luxurious comforts that await him at the McHenry House.

OTHER PENNSYLVANIA OIL TERRITORY.

In the foregoing pages a full description has been given of the oil region of Venango county, Pennsylvania. There are, however, several other localities in the State where oil has either been proved or is supposed to exist, in considerable amount. The Upper Alleghany and the Broken Straw Creek, in Warren county, have been selected by many oil seekers as desirable sites for investment, and numerous wells are going down on the banks of those streams and their tributaries. So far, no practical result has followed those experiments, although in many instances the "show" is good. In Crawford county, the two branches of Oil Creek and the upper portion of French Creek have been taken up by speculators, and derricks are going up in great numbers. In Erie county, wells have been sunk in the city of Erie, considerable gas and some oil being found.

The Clarion river, through Clarion, Elk, Forest, and Jefferson counties, has been the scene of great

(118)

excitement on account of the "oil strikes" reported from time to time. A large number of wells have been sunk along the Clarion and the streams running into it, several of them meeting with moderate success. As yet, the Clarion River territory can only be considered as "very promising," although the fact of the existence of oil on it has been thoroughly demonstrated. The Mahoning river in Lawrence county, and Slippery Rock Creek in Lawrence and Butler, are both considered good oil-producing territory. On the Slippery Rock about a dozen wells are down or nearly down to the proper level, and some are producing oil at a depth of six hundred to seven hundred feet.

South of Pittsburgh, on the Monongahela river, in Fayette and Greene counties, a number of experimental wells are going down. On Dunkard Creek, an affluent of the Monongahela, in Greene county, several wells have been sunk, with good success, and a considerable amount of oil has come into market from the Dunkard region. All Western Pennsylvania appears to have more or less oil beneath its surface.

THE WEST VIRGINIA OIL REGION.

FOLLOWING the line of the Alleghanies on its western slope, the oil region of Southwestern Pennsylvania appears to be continued into Northwestern Virginia. Indications of oil have been found in the "Pan Handle," and the counties immediately south of the Pennsylvania line, and some experimental wells have been sunk with moderate success. But the great oil territory of West Virginia lies on the Little Kanawha and Hughes Rivers and the numerous creeks and runs pouring into them, or into the Ohio, in Pleasants, Ritchie, Wirt, and Wood counties, and is known as the "Great Oil Belt." The geological formation of this territory differs entirely from that lying beyond its limits. The "upheaval" or "Oil Belt," extends from the Ohio River, opposite the little Muskingum and Duck Creek, about forty miles in a direction a little west of south, varying in width from three to ten, or perhaps fifteen miles. The rocks are peculiarly disturbed and broken ; the hills, along the numerous streams and gorges, varying from one to three hundred feet high; and along the centre of the belt the rocks are nearly vertical, but dip at various angles as they recede on either side, forming what is called the East and West slopes. By some convulson of nature, the rocks appear to have been "up-heaved," and separated, making deep ravines, gorges and gullies, many of which have become the permanent beds of streams, along the bottoms of which is found the "boring territory" as indicated by the color and character of the rocks, and the presence of oil, both on the surface and oozing from the fissures of the rocks.

The principal streams arising in, and running through these gorges or openings, and penetrating the Great Oil Belt, are the Little Kanawha, Hughes River, with its North and South Forks; Goose Creek, with its Laurel, Pigeon Roost, Oil Run, Myers, Ellis', Buffalo, and First and Second Big Run Forks; Mill-Site Run; Walker's Creek, with its Straight Walker Fork, Silver Run and Bee Tree Run Forks; Stillwell Creek, with several forks; Bull Creek, with its Horse Neck and Isaac Forks; Cow Creek; Calf Creek; Rawlson Run Fork; French Creek; Standing Stone Creek; Burning Spring Run, and other streams of less note.

This Oil Region is completely undeveloped, yet

11

the existence of Petroleum or British Oil, as it was
called by the settlers, has been known for more
than fifty years. Thousands of barrels of oil
have been taken from pits sunk in the sand on
the banks of Hughes River. In 1860–61 the high
oil fever existing in the Venango Valley of
Pennsylvania, spread out to this region, and
several enterprising companies and individuals
commenced boring for oil, on the Hughes River
at Oil Springs, and on the Kanawha at Burning
Springs. At the former place a flowing well
was struck, which, up to to-day, continues to
flow at the same rate as when first opened, from
two to six barrels of oil per day. At the Burning Springs, the great Llewellyn well was struck,
which flowed for several months at the rate of
from fourteen hundred to two thousand barrels
per day, and is still largely productive. A great
many wells were commenced at different localities; some on Cow Creek, Stillwell, Oil Creek,
Walker's Creek, and the creeks near the Burning
Springs, all of which that were not broken up by
the Rebels at the beginning of the Rebellion, produced oil in greater or less quantities. The fifteen
or sixteen wells on Oil Creek, at Petroleum,
yielded, and still continue to yield, from two to
three hundred barrels of superior lubricating oil
per month, their depth being only from eighty
to one hundred and sixty feet.

On account of the rebellion, all operations were suspended from 1861, up to Sheridan's successes in 1864. Since then this region has been the theatre of the most intense excitement. Experienced oil men from Pennsylvania have secured large tracts of oil lands; numerous enterprising companies have been formed, securing hundreds of acres of the choicest boring territory, and to-day, where but recently all was so still and desolate, may be seen nearly three hundred derricks, and the cheerful puffing of as many engines.

In visiting the oil regions of West Virginia, the first point to be made is Parkersburg, which is to the West Virginia oil territory what Oil City is to that of Venango county — the great *entrepot* and shipping point. In coming from the East the best route is by rail to Pittsburgh, and thence by steamer to Parkersburg; or by rail via Pittsburgh or Cleveland, to Wheeling, and thence by steamer to Parkersburg. From the West and South steamers can be taken from Cincinnati. Baltimore passengers can take the Baltimore and Ohio Railroad to Grafton, and go thence by Northwestern Virginia Railroad directly through the heart of the oil region to Parkersburg.

The principal producing oil territory on the Little Kanawha, is Burning Spring Creek, on which was the great Llewellyn and other flowing wells of the past.   To reach this territory the oil-seeker should secure a horse in Parkersburg, the distance being about thirty miles.   The road passes through Elizabethtown, which in 1861 was the scene of a great oil excitement, thousands of people being there " prospecting."   The breaking out of the war scattered the oil-seekers, and Elizabethtown had other, and far more unwelcome visitors, in the shape of guerrillas, who hunted up unfortunate Union men instead of oil springs, and " prospected " for plunder rather than for well sites.

Burning Springs Creek enters the Little Kanawha from the North.   The line of the Creek is marked by evidences that it was once a scene of successful oil enterprise, and that attention has again been directed to it.   Besides the Llewellyn well, already mentioned, there were several other flowing wells that produced largely, and some of them are still yielding.   The Rathbones owned a number of wells at Burning Springs, and lost heavily when the wells, with twenty thousand barrels of oil, were burned by the rebel guerrillas. The old " Eternal Centre" well still flows intermittently, yielding about twenty-five barrels a day.   There are some pumping wells, yielding from twenty to a hundred barrels.   About fifteen miles up the Little Kanawha from Parkersburg the Hughes River enters from the East, dividing, a few miles above its point of debouchment, into the North and South Forks. Near the junction of the forks are several oil springs and oil wells. The land on both forks has been extensively taken up by Eastern companies for the purpose of development. The surface along the route of the Hughes River is very much broken. Many years ago the petroleum was obtained from the banks of the river by digging shallow pits, which still are plainly visible. It is said several thousand barrels were thus obtained. Several wells are putting down, but none of them have reached a sufficient depth to properly test the territory.

Goose Creek enters the Hughes River from the North, and is crossed by the railroad at Petroleum Station. Near this point it is entered by Oil Run, on which there are a number of shallow wells. Fifteen wells, known as the "Petroleum Wells," are pumped with one engine.

On the South Fork of Hughes River, near McFarland's Run, is an immense bed of what is known as "Petroleum coal," resembling in its general features the famous "Albert coal," of Nova Scotia. It neither resembles coal nor asphaltum in appearance or properties. It is found in a "lode" or vein, running nearly East and West. An analysis proved that it yielded by distillation one hundred and sixty-nine gallons

11*

of crude oil to the ton, which, on refining, loses only fifteen per cent.

North of Parkersburg, and higher up the Ohio River than Marietta, are Bull, Cow, Calf, Crow, and French Creeks. On the latter, and most northern, are several wells going down. On Crow Creek is one well producing much gas. Cow and Calf Creeks are also being thoroughly bored. Bull Creek, about thirty miles above Parkersburg, is now a favorite oil locality, and numerous wells are going down. About six miles above its mouth, on Horse Neck Run, is the celebrated Horse Neck well, which has produced a large amount of oil. On Rawson's Run, which connects with Horse Neck Run, are several wells owned by the American Oil Company, and Messrs. Tack & Brasher, which have been very successful.

It is a fact, worthy of consideration, that not a single well has been abandoned on account of the failure or non-appearance of oil, while in almost all other oil regions, a large number have been abandoned as "dry wells." It is believed that no greater or more productive wells have been opened in any region, than in West Virginia, and in proportion to the time spent and capital invested, nowhere has the enterprising oil-seeker found a more sure and abundant return. The development in Pennsylvania began about ten years ago, a vast amount of capital has been ex-

pended, and nothing has occurred to retard the
vigorous prosecution of explorations. The success there has been very great, almost fabulous.
In West Virginia, the enterprise is but just begun,
yet all experienced oil men, and the most skilful
geologists, concur in the belief that the same
time and similar enterprise will develop here an
equal, if not a greater yield of oil. At Horse
Neck there are some twenty wells, producing
from ten to sixty barrels per day. D. H. Wallace,
in company with the Phillipses, of Oil Creek,
own about five thousand acres of oil territory on
and in the vicinity of Bull, Cow, and Calf Creeks,
and have some sixty wells in operation, or in
process of boring, with suitable engines. There
are several wells going down on Stillwell; one on
Walker's Creek, by Mr. Murray, near the Smith
Farm, having a fine show of oil at the depth of
one hundred and sixty feet, and having at thirty
feet gone through a stratum of copper ore, of
superior quality, about thirteen feet thick. Mr.
Candy has two wells in progress above the Petroleum Wells. The Great Belt Oil Company, of
Cleveland, have one well down over three hundred feet on the Hall Farm, and one two hundred
and sixty feet on the Sharpnack Farm, near the
"Oil Springs," with oil sufficient to warrant
tubing. The wells on Oil Creek, known as the
Petroleum Wells, are doing finely, considering

the fact that fifteen of them are pumped by one engine. That company is putting down several new wells, with every show of success.

The Baltimore Company, Mr. Cannon, President, are prosecuting their works with becoming energy near the Oil Springs, on Mill-side Run, just below the Sharpnack Farm of the Great Belt Oil Company, where they have one well down over five hundred feet with every prospect of success, and two more engines and derricks up and nearly ready for work. The wells in the vicinity of Burning Springs are all making a good yield, and next spring a large number of wells will be put down on the Rathbone tracts and on the Standing Stone, by several enterprising companies.

From the number of working companies formed, it is safe to predict that during the coming year rich developments will be made, and great wealth accumulated in the region of the Great Oil Belt of West Virginia.

The oil found in West Virginia is of superior quality. Most of the shallow wells produce a lubricating oil, very heavy and of great value, the illuminating oil having none of the offensive odor that is sometimes found; it is of excellent quality for refining. The lubricating oil commands readily thirty dollars at the wells. Geologists and experienced oil men concur in the belief that the illuminating oils will be found by sinking the wells producing the lubricating oil to or through the third sandstone.

So far as has yet been discovered by boring, the description and strata of rocks in the West Virginia region are the same as those in Venango county, Pennsylvania. The upheaval is composed of a reddish brown sandstone. The strata below, as found by boring in the valleys of the streams, are about as follows in both regions, varying somewhat as the break or the slopes of the Belt are approached. First sandstone, from thirty to two hundred feet; soft rock or shale, from ten to one hundred feet; second sandstone, from fifty to one hundred and fifty feet; shale or soapstone, ten to thirty feet; third sandstone, from sixty to one hundred feet; shale or soapstone, from thirty to fifty feet; fourth sandstone, from sixty to one hundred feet; limestone.

I

SOUTHERN OHIO OIL REGION.

DIRECTLY across the Ohio River, in a line with the West Virginia Oil Belt, is a productive oil territory, embracing Washington, Meigs, Athens, Morgan, and Noble counties. The business centre of the district is Marietta, which can be reached by rail from Cincinnati, or by steamer from any point on the Ohio River. The locations of oil enterprise in this region are the Muskingum and Little Muskingum Rivers, Federal Creek, Wolf Creek, and Duck Creek.

Duck Creek enters the Ohio about a mile above Marietta, which is at the mouth of the Muskingum. Operations commenced on this Creek in 1860 during the first oil excitement in Pennsylvania, and several wells were sunk to a depth of from one hundred to two hundred feet, and oil in sufficient quantity obtained to warrant pumping. The low price to which oil fell, and the troubles incident to the breaking out of the war, led to the abandonment of all the wells, and the district remained

(130)

undisturbed by oil seekers until a few months since. Taking a horse at Marietta and riding across the country to the lower part of Duck Creek, the visitor soon comes in sight of the derricks, old and new, which line the lower part of the Creek. The work, past and present, in that locality, is of a very primitive character, all the boring having been done by the "kicking" process, engines being considered too costly. The wells are therefore shallow, yielding but a small quantity of oil, although of heavy quality.

About ten miles from Marietta, near Lower Salem, the first evidences of active and business-like oil enterprise become visible. On the Paw-Paw, a small stream flowing into Duck Creek, a number of wells are going down, some of them having already met with fair success. But it is not until reaching Macksburg, about ten miles higher up, that the most valuable part of the Duck Creek region is reached. Here there are several important wells, down to a depth varying from eight hundred to eighteen hundred feet, and yielding largely. Among the more noticeable are the Dixon, Dutton, and Steel wells, all of which have proved highly valuable. For several miles above Macksburg, Duck Creek is lined with derricks, over fifty engines being hard at work. The surface of the water is covered with floating oil, and the whole appearance is strongly suggestive of Oil Creek. The product of the Duck Creek region, around and above Macksburg is taken across the country, by villainous roads, to Lowell, on the Muskingum, about ten miles distant, and thence boated down to Marietta.

Passing up the Muskingum to McConnellsville, in Morgan county, another oil region is reached. Here the surface indications are unusually rich, and several shallow wells have been successfully worked. The territory has been extensively leased, and some of the companies propose to sink deep wells for the purpose of developing the value of the lower deposits.

Wolf Creek enters the Muskingum from the South, about midway between McConnellsville and Marietta. The oil territory can be reached by a ride or walk of five miles from the former place. Most of the oil has been obtained at the mouth of Buck Run, on Wolf Creek. The older wells are shallow, finding oil from seventy to one hundred and twenty feet. Several thousand barrels have been obtained from this neighborhood. The "Deep Oil Well" was sunk three hundred and eighty-five feet, and the memoranda of the boring, as furnished by the manager were as follows:

"Found the fossil rock twenty-seven feet below the surface. The first oil vein was struck at the depth of fifty-three feet, from which were obtained thirty five barrels; second oil-vein one hundred feet from the surface, showing more oil than the first, but it was not pumped; third vein, two hundred and seventy-five feet, in black shale bituminous — considerable oil, not pumped; fourth vein, three hundred and forty-five feet deep, quite a good show in the blue sand-rock — oil all the way through that rock, which was thirty-five feet thick; fifth vein was in a blue-black shale. Oil was found through this shale, three feet. Sixth vein in a blue sand-rock, three hundred and seventy-eight feet deep, and within six feet of the bottom of the well. Ceased boring at three hundred and eighty-five feet, two feet in a white sand-rock, called the 'salt-rock.' Pumped three days and nights, when, by the breaking of the seed-bag and pump, after obtaining about ten barrels of oil, operation ceased."

Keeping west from Wolf Creek, the oil territory of Federal Creek, in Athens and Morgan counties is reached. Here the surface indications are also very rich, and the yield from shallow wells has been large, much greater than from any other locality of the State. The utmost activity is displayed in the work of developing the territory, twenty engines being now at work, and preparations making for more. The most celebrated location is the Joy Farm, on which thirteen wells are down or preparing to go down. The oil obtained here has been from a depth of

12

forty to one hundred and twenty feet. Three deep wells are going down on the Creek, for the purpose of testing the question of the existence of " deep oil." Federal Creek enters the Hocking River at Federalton.

These comprise the more important portions of the Southern Ohio oil region. Indications are found in counties adjoining, but as yet the excitement is purely speculative, no actual tests of the value of the deposits having been made. In Scioto, Pike, and Ross counties, along the Scioto River, and especially in Pike county, near Jasper, very heavy petroleum or "mineral tar" distils from the bituminous shale which there crops out to a thickness of two hundred and fifty feet. The same manifestations are visible in Adams county, on the Ohio, just below the mouth of the Scioto. The rock is so highly charged with petroleum that it burns readily, and a large percentage of pure petroleum can be extracted from it. Companies have been formed for buying the shale rock and distilling the oil from it.

NORTHERN AND EASTERN OHIO OIL REGIONS.

THE oil fields of Butler and Lawrence counties in Pennsylvania, which have lately come into notice, are continued across the Ohio line into Columbiana, Mahoning, and Trumbull counties, following the line of the Big and Little Beavers, the Mahoning, and other tributary streams. Along the Big Beaver and Mahoning there are a number of wells, many of them yielding moderately, but none to any great extent. The wells are mostly shallow, especially towards the West. Experimental wells are going down as far up as Warren in Trumbull county. On a small tributary that enters the Mahoning from the north is the well-known Mecca district that in 1860 and 1861 was the scene of considerable activity. The wells are shallow, averaging fifty feet, and yield a very heavy lubricating oil, worth more than six times the value of the Pennsylvania illuminating oils. The yield of the wells is very small, four barrels being an extraordinary amount.

In Columbiana county, on the Little Beaver, is the Smith's Ferry district, also the scene of considerable excitement in 1861, when a number of wells were sunk, with but moderate success. The tide of oil enterprise has again set towards this region, and over twenty wells are in operation, some of them obtaining from one to eight barrels of oil per day.

The Rocky, Black, and Vermilion rivers entering Lake Erie from Cuyahoga, Lorain, and Medina counties, have also become the scene of oil enterprise, and a number of wells have been put down. None have as yet reached the proper depth. In 1860–61 several wells were sunk in the vicinity of Liverpool, Medina county, and a small yield of heavy lubricating oil, similar to that of Mecca, obtained at a depth of one hundred to two hundred feet.

THE OIL REGIONS OF KENTUCKY.

THERE are four localities in Kentucky that have come into notice as oil territory, namely: the Big Sandy, on the Eastern border, and forming part of the Great Kanawha and Guyandotte Basin of West Virginia; the Cumberland River district in the South, extending into Tennessee; the Vanceburg district in Lewis county, opposite Adams and Scioto counties in Ohio; and the Green River district, in McLean county, Western Kentucky. Of these the best known, and, at present, the most developed, is the Big Sandy district.

The point of departure is usually Catlettsburg, at the junction of the Big Sandy and the Ohio. Steamboats run up to Louisa, a distance of about twenty miles, and from that point there is generally good skiff navigation, as well as walking or horseback facilities. Blaine's Creek enters the Sandy below Louisa, and at the latter point the river divides into West Fork and Tug Fork. The river

12*   (137)

and its tributaries have long been known as productive of oil, the surface of the streams being frequently discolored with patches of floating oil, arising from oil springs in the bed of the stream and in the pools adjacent. For many years this surface oil was gathered in large quantities, and used for various domestic and medical purposes. In 1859 and 1860 a number of shallow wells were sunk along the valley, all being put down by hand, and reaching the depth of from one hundred to two hundred feet. None of them were pumped, the breaking out of the war putting an effectual stop to oil operations in Kentucky. In most of the wells bored, and in some old salt wells in the valley, the flow of oil into the well from veins cut, gave assurance of good returns on the investment, had the condition of the country allowed the prosecution of the enterprise. Near the Big Burning Springs, on a branch of the Licking River, heading near Paintville in the Sandy Valley, a well was bored to the depth of one hundred and seventy-three feet. A column of oil and gas was thrown up and continued to flow for days. Before the necessary machinery could be procured to work the well, the war commenced, and the well was left to blow itself out.

In the vicinity of Paintville, near the junction of Paint Creek with the West Fork of the Big Sandy several wells were put down in 1861 with

fair prospect of success had the work been continued. The land in the neighborhood has now
been leased by Eastern and Northern capitalists,
and the preliminary work for sinking numerous
wells has been commenced. The whole line of
the river and its tributaries has been "prospected"
by speculators, up as far as it is safe to go for the
guerillas who still prowl around Southwestern
Virginia and the Kentucky border.

The Cumberland River district is on the river
in Cumberland county, near the Tennessee line.
The whole valley for many miles is rich in indications of oil, and for many years the existence of
oil has been known in the neighborhood. About
three miles above Burksville, in Cumberland
county, is the celebrated "American Oil Well,"
the history of which is narrated in *Niles's Register*
for 1829. From that account we learn that
during that year whilst some men were boring
for salt water, and after penetrating about one
hundred and seventy-five feet through a solid
rock, they struck a vein of oil which suddenly
spouted up to the height of fifty feet above the
surface. The stream was so abundant and of
such force as to continue to throw up the oil to
the same height for several days. The oil thus
thrown out ran into the Cumberland River, covering the surface of the river for several miles. It
was readily supposed to be inflammable, and upon

its being ignited it presented the novel and magnificent spectacle of a "*river on fire*," the flames
literally covering the whole surface for miles,
reaching to the top of the tallest trees on the
banks of the river, and continued burning until
the supply of oil was exhausted. The workmen
abandoned the well in disgust.

In 1860 four firms from Virginia came into the
neighborhood of the old well, and commenced
boring for oil. Fine shows of oil were obtained,
and the work of pumping was ready to be commenced, when the rebellion broke out, and the
wells were abandoned. Some of the parties engaged in the enterprise in 1860, have returned to
the spot, and recommenced work. They have
been followed by others, and preparations are
making for a vigorous summer campaign. The
oil of this district is about 35° in density.

Burksville is about two hundred and fifty miles
above Nashville, has good steamboat navigation
for boats of five hundred to six hundred tons
burden for half the year and flat-boat navigation
for the remainder of the year. To visitors from
the North the nearest railroad point is Horse
Cave, 45 miles distant, on the Louisville and Nashville Railroad. The present travelled route, from
Louisville, is via. Louisville and Nashville Railroad, eighty-three miles to Cave City, stage twelve
miles to Glasgow, and thence forty miles, on horse-

back, to Burksville. Another route is by rail, fifty
miles to Lebanon, by stage forty miles over a turn-
pike to Columbia, and thence thirty miles on horseback.

The Green River district has hitherto met with
but little attention, but a movement has now set
strongly in that direction. In 1861 a well was
put down near Calhoun, McLean county, to the
depth of one hundred and thirty feet, and abandoned on the approach of the rebel forces. Louisville parties have now taken the old well and are
pumping about two barrels of heavy lubricating
oil daily. A number of mineral tar springs are
scattered along the line of Green River.

The Lewis county district is a continuation of
the oil region of Adams, Scioto, Pike, and Ross
counties in Ohio, and the manifestations are similar. Bituminous shale, heavily charged with thick
petroleum, crops out, and considerable deposits of
thick oil have been found. About fifty engines
are working, or being set up, in the neighborhood
of Vanceburg.

OIL IN OTHER STATES.

CONSIDERABLE excitement on the subject of oil has broken out in Indiana, and large tracts of land have been taken up for the purpose of boring. As yet, no producing wells exist in the State, sufficient time for properly testing the value of the lands as oil territory not having elapsed. The presumed oil regions of Indiana are situated at some distance from each other. The most favorably located region is in the southern part of the State, bordering on the Ohio River, and being apparently a continuation of the Green River district of Kentucky. The counties included in this district are Harrison, Dubois, Martin, Perry, Crawford, and Orange.

In Northwestern Indiana no attempts have yet been made to test the alleged existence of oil in considerable quantity, the excitement existing in Jasper, Newton, Benton, and Fountain counties, being purely speculative. Whether oil exists there to any extent remains to be proved.

Indications of oil exist in the States of Illinois,

Iowa, Missouri, and Michigan, and some attempts have been made to test the productiveness of the supposed deposits. At the present time no satisfactory results have been reached, the experiments not having had sufficient time for proper development.

In Western New York the indications of oil are very promising. In Alleghany, Cattaraugus, and Chautauqua counties, a number of experimental wells are going down, with variable prospects of success.

Oil is also reported in Eastern New York and Connecticut; in Oregon, Utah, and California. In Canada it has been produced for three or four years, and lately it has been reported in Mexico. There appears to be as little limit to its field of production as there is to its usefulness.

INVESTMENTS IN OIL WELLS.

In the history of the commercial and industrial interests of the world there is to be found no complete parallel to the circumstances connected with the discovery of petroleum and the development of its uses. Other important agents have been discovered, but their value was not readily perceived, and their extensive adoption has been the result of long years of experiment. Six years more than covers the entire history of petroleum, so far as the industrial world is concerned. Previous to that time it had slumbered deep in earth, floated unnoticed on the surface of stagnant pools, or dripped unheeded from the rock. Now it has become an article of necessity everywhere. It forms one of the great staples of commerce. New branches of manufacture have been created by its aid, and those already in existence have received additional impetus. It furnishes employment to the workman and cheerful light to his home. Thousands of wheels that otherwise would have remained idle, have been set in motion, and tens of thousands of busy arms kept employed by its agency. Since the discovery of the uses of steam no event of similar importance has occurred in the industrial world. But the introduction of steam into general use was a work of many years. Prejudices and motives of interest were opponents against which the new-born power had to wage a long and fierce contest before its rights were acknowledged.

Petroleum had no such opposition to contend with. The want existed before the means of supplying it was discovered. The supply of fish and animal oils was yearly diminishing, whilst the demand was yearly increasing. The question as to the source from which the illuminating and lubricating agents were henceforth to be obtained had become so important as to tax the minds of the leading scientific men in the search for a satisfactory answer. At that moment petroleum became known and its value was immediately investigated. For every purpose in which fish, animal, and most of the vegetable oils were used, petroleum was found more serviceable and incomparably cheaper. Its adoption for those purposes immediately followed, and experiments proving its possession of valuable properties not found in other oils, the demand became proportionately greater, far exceeding the supply.

The extraordinary success of the pioneers in oil mining, the immense returns frequently resulting from small expenditures, and the high value of the product, created by the rapidly increasing demand, soon turned public attention to the subject, and capital was freely invested in oil enterprises. Some of these were failures, but the prizes won were so enormous in proportion to the stakes that the occasional disasters were wholly lost sight of, or served but as warnings against some particular location or mode of working. The manifest folly of risking the entire means of an individual on the hazard of a single well yielding oil led to the establishment of joint-stock companies for the prosecution of the enterprise, and the eagerness with which their shares were taken up soon multiplied these companies to an enormous extent. In the Spring of 1864 several new companies were organized. During the Summer and Autumn their increase was rapid, and in the Winter of that year, and Spring of 1865, new companies were organized daily in the leading cities. To give their present number would be useless, when every day brings an addition.

Of course very many of these companies have no real value. Their lands, where any title to lands really exists, have no indication of the presence of oil in quantities to warrant boring. The only object of their existence was the creation of shares to be sold at a profit by the sharp-witted projectors. The originators of such companies are moral swindlers, and only evade the legal responsibilities of actual swindling by the ingenuity with which their "prospectuses" are framed. On close examination it will be found that, although apparently asserting the value of their lands as oil territory and promising that rich results will follow their working, they in reality assert nothing and promise nothing. They "keep the word of promise to our ear, and break it to our hope."

Of the remaining companies, organized in good faith, a large proportion will probable meet either with failure or but small success. It can scarcely be otherwise when it is considered that in their anxiety to get possession of lands in which oil is supposed to exist, many of these companies have been compelled to purchase partially or wholly undeveloped property, the sole inducement being the existence of one or two good wells on the same stream, although several miles distant. Some companies will, as in many previous instances, prove mines of wealth, whilst others will yield a good business percentage on the investment.

The "oil fever" that at present pervades all ranks and classes in the United States, and is spreading to other lands, has been compared to the South Sea Scheme, Law's Mississippi bubble, and the *Morus Multicaulis* excitement. It differs from all of these in the important particular that it is based on an actual fact, of proportions so gigantic that it serves as an excuse for the most extravagant anticipations of those interested. The cases cited were essentially bubbles, wholly speculative and based on a supposititious state of affairs. With the first breath of distrust the bubble burst, and not a vestige was left of the magnificent schemes save the ruined fortunes of their believers. That the present inflation will, sooner or later, be followed by a corresponding collapse, must be expected. Hundreds of companies will go down beneath the crash, and sink in merited oblivion. With them will go many enterprises deserving a better fate, but which will be unable to stem the downward tide. Even those which are well established and have given proof of their solvency and reliability will probably suffer for a time, but in the end will rise to their legitimate position.

The only instance in the speculative excitements of the past at all resembling the "oil fever" of the present day was the English "railway mania" of 1845. The success of the railways then existing stimulated investment in similar enterprises to such an extent as to create a species of insanity for the formation of railway companies and the possession of railway shares. Lines of the most absurd and impossible character were projected and found ready sale for their shares. Of course the delirium ended at last, and with returning sanity came the extinction of hundreds of companies and the ruin of thousands of people. But the real value and importance of railways remained unimpaired, for there was a solid basis on which the speculative superstructure had been erected, and the past twenty years have seen the rapid and permanent growth of a railway system but little inferior in extent to the wild dreams of 1845.

That this will be the case with the petroleum interests of the United States cannot be doubted. The business which in six years has grown from nothingness to its present gigantic proportions cannot be seriously injured even by a great panic, should such an event occur. It has been woven so closely and thoroughly into our daily life that the demand must continue to increase. From Maine to California it lights our dwellings, lubricates our machinery, and is indispensible in numerous departments of arts, manufactures, and domestic life. To be deprived of it now would be setting us back a whole cycle of civilization. To doubt the increased sphere of its usefulness would be to lack faith in the progress of the world. To fear the exhaustion or diminution of

13*

its supply would be to doubt the beneficence of the Providence that never makes known a benefit to man for the purpose of distressing him by its withdrawal.

A noticeable feature in the history of the petroleum business of the United States is the peculiar period at which it made its appearance, and the important part it has played in our national affairs. At the time when we were struggling to maintain our existence at an enormous cost, when the balance of trade with Europe was against us, and the shipments of gold were draining the life-blood of our commerce, a sudden and unexpected trade sprung up in the article of petroleum. From a few thousand gallons in 1861 the annual exports from the United States increased in 1864 to nearly thirty-two million gallons, the value of the shipments for the past two years amounting to over thirty millon dollars, thus materially reducing the export of gold and increasing the prosperity of the country by this means, as well as by stimulating industry within the national borders. This was effected by the employment of superabundant capital and the development of lands that had hitherto been considered valueless. To facilitate the work railways have been built and are building, which would not have been constructed but for the existence of petroleum, but which will prove of great value in opening up neglected territory, enhancing the value of lands, and at the same time be profitable investments in themselves.

A word of caution with regard to investments in oil lands and oil stocks will not be out of place. It is well to remember that " all is not gold that glitters," and the "good thing" which the plausible "projector" attempts to induce you to invest in, may be a good thing for *him*, but a bad thing for you. Before investing, see that the parties directly interested are responsible and honorable men. That is not all. Good and honorable men are sometimes prominently connected with enterprises of which they personally know nothing, trusting to the representations of some other person directly interested. Ascertain that point before risking the money you may never see again. With regard to the particular section of country in which to invest, no other advice can be given than that it is, as a general rule, safer to put money into property where paying wells already exist — should such property be in the market — than in wholly untried country. At the same time, where the real value of the property is unknown, the stakes are smaller, and the prizes, if found at all, proportionately larger. It by no means follows that because one well has proved a great success, another sunk in its neighborhood will also be fortunate, although the enormous increase in the value of land for miles around a successful well immediately after the "strike" seems to prove the generality of that belief. At the same time, there is better hope of obtaining oil when surrounded by yielding wells than if encompassed by "dry holes."

That considerable sharp practice is frequently connected with the formation of oil companies and disposition of the shares is to be supposed. Sharp-witted men do not work for nothing, and such men have found a rich harvest in the oil speculations of the present day. A man of this class goes into the oil territory, finds a patch of land suited for his purpose, and engages to buy it at five or six times its real value, paying from twenty-five to one hundred dollars as deposit, and agreeing to pay the balance on a certain day, or forfeit his deposit. He then endeavors to dispose of it to a party of fellow speculators at twenty-five to fifty per cent. advance. These again put it into a stock company of, say, $100,000, "reserving $25,000 as working capital." The remaining $75,000 of stock is divided among the "partners on first principles," as pay for the land, so that if the stock sells freely at a fair figure, a handsome profit is divided. It is a common practice for the projecters of a company to connect a large amount of undeveloped land with an interest in some dividend-paying well as "a sweetener." In some cases this proves a good thing for the stockholder, and in other cases he is "sweetened."

Frauds are not unfrequently perpetrated by "planting" oil in dry wells. Some notable instances of this have been made public, in which the contrivances were of the most elaborate character. The petroleum interest is no more accountable for such crimes than society is chargeable with the guilt of the occasional pickpocket or burglar.

In conclusion, invest no more money in oil-wells, or in any other speculative business, than you can lose without being crippled in resources If able to spare the means, examine closely, judge carefully, then invest boldly and await the result with patience, with our wishes that you may "strike oil."

THE END.